FUNGI
Delight of Curiosity

Cyathus olla, one of the Bird's Nest Fungi, magnified about four times. This pea-sized fungus, related to the puff-balls, is often found on garden soil. Its little cups, a marvel of the fungus world, are beautifully designed splash 'guns,' from which the seed-like spore-cases are 'fired' for reproduction by the weight of splashing raindrops.

FUNGI

Delight of Curiosity

Harold J. Brodie

UNIVERSITY OF TORONTO PRESS

Toronto / Buffalo / London

©University of Toronto Press 1978
Toronto   Buffalo   London
Printed in Canada
Reprinted 1979

Library of Congress Cataloging in Publication Data

Brodie, Harold Johnston, 1907–
Fungi, delight of curiosity.

Bibliography: p.
1. Fungi. I. Title.
QK603. B85   589'.2   77-25204
ISBN 0-8020-2289-8

Publication of this book has been assisted by the Canada Council
under its block grant program

Contents

Illustrations

Preface

It has been my wish in the writing of these pages to share with others the exhilaration which I have always experienced in the presence of those living organisms which captivate attention because of the colour, the form, and the almost incredible nicety and precision of their structures and complex functions. Persons who already possess some awareness of the beauty of our natural environment will, I hope, derive pleasure from the reading, if they also possess sufficient curiosity so that they wish to understand more than is afforded by cursory inspection. These pages contain nothing new for biologists, who are already fully familiar with the facts that are here presented. Moreover, these pages are not intended to be instructive according to the standards of technical detail and objectivity which biologists (all scientists, indeed) properly expect and require in the reports to which they are accustomed to give serious consideration.

Sharing of experience may be, and often is, of a very simple nature, as when two good friends wander through a woodland in summer; one comes upon an exquisite wild orchid and calls to the other to share the sight. Even if there is nothing more, the simple process of sharing the experience greatly enhances its significance for each observer, but the experience itself remains nebulous and undefined. Deeper understanding and more enduring appreciation require the definition that comes only from something more than casual attention.

Preface

Unfortunately, or so I believe, the potential of the impression that can result from the first sight of an unusual natural object is seldom developed fully. Enthusiasm is easily dampened by the mere effort of learning the name, scientific or common, of the object in question. In one category are many persons whose curiosity seems to be sated as soon as they have learned the name; they write the name conscientiously in their notebooks and give the plant or other natural object no further attention, apparently quite happy to proceed to another plant in search of yet another name. In another category are those who are discouraged because the name, when they do find it, somehow destroys their illusions. The Latin and correct scientific name of the wild orchid, *Orchis rotundifolia*, immediately becomes for many persons an obstacle which mysteriously prevents full enjoyment of the lovely flower. The name in English, 'Round-leaved Orchid' (simply a translation of the Latin), might be more acceptable; but when another commonly used English name comes up, namely 'Fly-spotted Orchid,' the allusion to the activity of flies seems to be utterly out of place when connected with the floral jewel.

And there are other even more effective dampers of initial enthusiasm, the principal one being the specialized, technical nature of the descriptions and other kinds of printed records dealing with the organism. Mere lack of acquaintance with highly specific technical terms, by itself, commonly occasions discouragement; and even if this difficulty is overcome, there remains the matter of idiom. The language of science is in large measure, though by no means exclusively, didactic, unequivocal, and factual; it is not the language of poets and philosophers. What I have wished to do is to reassure the reader that 'rosy-fingered dawn' will return again and again, and still rosy-fingered, long after we have become involved in understanding the physical phenomena that underlie the production of the sunrise or sunset colours. Likewise, fascination with the lively green colour and neat microscopic brushes of the common mould *Penicillium* does not diminish as soon as we have learned that it produces a life-saving drug, penicillin.

'Beauty is in the eye of the beholder' is a well-worn phrase which, like many other enduring things, need not be discarded because time and frequent use may have rounded off its edges; time and use have also added to its lustre, for they have repeatedly proved its worth as a statement of truth. Without contesting the prerogative of the reader to view the fungi or any other natural object from any vantage point that may be desired, I have nevertheless attempted to show that even a slight shift in viewpoint may very easily enhance enormously the pleasure and intellectual stimulus which anyone can derive from contemplation of many fungi to be found at our own back doors – and at no expense, no inconvenience, and with only a modicum of effort in directing the faculties with which we are blessed.

In our times, much of man's mechanical world, as well as his social world, produces feelings of instability and uncertainty. For this reason alone there may be pleasure in contemplating a stable world, even though what is here presented is only a small part of a microscopic realm. Living organisms change, of course, but they usually do so imperceptibly slowly; there is no evidence whatever to suggest that the appearance of the Coral Fungus (Figure 10) is any different today from when Fries first saw it or that the Bird's Nest Fungus (see frontispiece) has changed perceptibly since Clusius first recorded it in 1601. And we should note that Clusius did not see much, because his viewpoint was restricted not only by lack of a microscope but also by the limitations of background knowledge in his time.

For readers who wish to delve deeper into topics presented only sketchily in this book, a short list of more detailed but not highly specialized books is presented. A brief glossary is also appended for the assistance of general readers. For those photographs which were not taken by myself, I am grateful to friends and colleagues

Preface

who generously provided them and whose names appear in acknowledgment.

ACKNOWLEDGMENTS

For permission to reproduce those photographs which were not taken by myself, the following acknowledgments are gratefully made; Figures 7, 8, 10, the late Dr W.C. McCalla (Calgary, Alberta); Figure 9, Dr David A. Pegler and Dr Tom W.K. Young, Chelsea College of Science and Technology; Figure 14, Dr George Barron, University of Guelph, Ontario; Figures 16 and 17, Dr R.K. Benjamin, Rancho Santa Ana Botanical Garden, California (Figure 17 was published in *Aliso* VII, p 173, 1970); Figures 18 and 19, Drs T.C. Vanterpool and Ruth Macrae, published in the *Canadian Journal of Botany* XXIX, p 148, 1951; Figure 20, Dr Neal Weber, Florida State University, published in *Gardening Ants, the Attines* (Philadelphia: American Philosophical Society 1972); Figures 15 and 21, Mr H.F. Dietrich, University of Victoria; Figure 5, Dr John Paden, University of Victoria.

Special thanks are also extended to Mr Dietrich (University of Victoria) for preparing prints suitable for publication of all the photographs.

For her patience and encouragement throughout the preparation of this book, I am deeply grateful to my wife, Mary.

HAROLD J. BRODIE

Victoria, British Columbia

FUNGI
Delight of Curiosity

1 *Botryosporium:* The beauty of a frost-covered shrub at night remains in the mind's eye, but the 'shrub' is a lowly mould, growing on a rotting cabbage leaf. Magnified about 50 times

Preamble

In the fable concerning the three blind men and the elephant, each man developed his own mind picture according to the nature of his particular contact with the elephant. The blind man who touched the sinuous and writhing trunk thought of a snake; he who took hold of the firm tough tail thought of a rope; and he who felt the rough massive leg pictured the base of a tree.

Like many fables, that story states very simply a truth which, in more extended form, leads far into the realms of philosophy and psychology. The reaction of each of us to the animate and inanimate objects of our environment (which Wordsworth called 'the influence of natural objects') is a personal thing; it is compounded of experience, training, sensitivity, and much more.

The pages that follow present a series of 'elephants' in the form of plants called fungi. Each object is examined first with the aid of what is admittedly a poor substitute for real life, namely a photograph. It is then examined by one of the blind men who can convey to you only those ideas that originated in his own experience, ideas that are obviously limited by his capacity for perception and by the difficulty of avoiding excessive technical language and unfamiliar concepts.

The name *fungus** is applied to members of a group of lowly living things more familiar under such names as mushrooms, mor-

---

* Technical words such as fungus are italicized on first appearance in the text and explained in the glossary.

els, truffles, and moulds. As natural objects having various degrees of interest and importance for mankind, they have been illustrated and described from the earliest days of printing; indeed, long before the advent of printing they were illustrated in carving, sculpture, and fresco. It follows that the pictorial and written records dealing with fungi constitute a substantial proportion of our knowledge of the living world.

The fungus plants that will appear in the book have been described, drawn, and photographed so frequently that the number of times is unimportant; the nature of these records is more meaningful than their number. Even a very few examples will show how varied and, more important, how significant has been the role of fungi in the adventures of the human race – and so it will ever be!

To the ancients, mushrooms were often symbolic of the supernatural; this may have been, as Mr Wasson has suggested, because primitive man long ago discovered that certain kinds when eaten can produce fantastic dreams. The ancients had their own notions about the origin of fungus growths; myths and sagas are replete with reference to fungi as the bizarre and unwholesome offspring from the union of the 'maleness' of a lightning bolt with the 'femaleness' of the earth.

By the time of the Romans, many common kinds of mushrooms were much esteemed as articles of diet, not so much for their intrinsic food value as for their unique flavour; and for the past two thousand years, fungi have continued to be prized as food with almost universal popularity. The thorns among these roses are the poisonous mushrooms; although many members of that fungus tribe can be eaten with impunity, a few contain deadly poisonous chemicals known as alkaloids. The records dealing with the frightful effects of fungus poisons do not make pleasant reading.

Many times in the distant and even not-too-distant past, fungi have assumed a very prominent role as 'historic determiners of human affairs,' as Professor Large puts it. That is because although many members of the fungus clan get their food from *dead* organic matter (such as old wood, leaves, and the like), other kinds are parasites that grow only (or preferably) upon *living* green plants, and some upon living animals. Were the fungi to parasitize, and so to destroy only thistles, dandelions, fleas, and mosquitoes, man would regard them as his allies; even then, however, so delicate is balance in nature that such an event might not prove a boon in the last analysis. But in all life there is competition for food and man competes with fungi in this respect, as he does with many other organisms among his fellow beings. Man cannot afford to share his table too liberally with rusts and other voracious parasites when they destroy his cereal crops, his vegetables, and his fruits. In fact, there is scarcely a plant which man may covet for his next meal on which some fungus may not have staked the first food claim. A microscopic mould, named *Phytophthora infestans*, decimated the Irish potato crop in the nineteenth century and thousands of Irish people starved; of those who survived, hundreds emigrated from Ireland to North America. This sort of struggle is not new; the Bible records the tribulations of the peoples of Asia Minor when blight and mildew ruined crops, and there can be little doubt that fungi had been taking their toll for thousands of years before biblical times. The same struggle goes on today, diminished somewhat on the one hand by the improvements in control methods made possible by modern science and aggravated greatly on the other hand by enormously increased crop production which exposes more food to the basic needs of the ravenous hoards of fungus parasites.

That which is fully exposed to view is often most effectively hidden. How could the wretched peasants and ignorant burghers of medieval Europe have known that the dread 'St Anthony's Fire' which burned as a horrible holocaust of human misery so often in their midst was caused by a small purple-black fungus that any one of them could have seen on heads of rye in their fields? Not seeing the fungus and in ignorance, they made the rye into their black bread and suffered the tortures of the damned because the purple-black rye fungus (we call it *ergot*) contained alkaloids that, when consumed, affect the human nervous system, musculature, and bones. Hundreds of pamphlets and several books have been devoted to description and illustration of this one parasite alone – for a fairly obvious reason.

Experts state that in many areas of the world fungi destroy annually more living trees, more lumber, and more finished wood products than does fire. Here again, because of the enormous economic implications, wood-rotting fungi form the subject matter of hundreds of books and thousands of bulky reports covering extensive and costly research projects.

And even man's own person is not free from possible attack. There are legions of medical research workers and practitioners who do little else but try to alleviate the suffering caused by certain moulds that are parasitic upon the human body, affecting mostly the skin; some of these parasites, alas, affect vital internal organs and result in death.

It is of course easy to over-emphasize the role played by fungi as man's enemies and competitors, so let us consider for a moment the credit side of the picture. Human beings have more blessings than they are generally aware of – among these, many fungi. It does not take much imagination to picture what would be happening if these organisms were not continually destroying or decomposing trees, leaves, and other vegetable matter. The bulky part of all vegetable matter consists of a great deal of a chemical material called *cellulose* and of chemically related substances. A few other minute organisms, such as bacteria and certain insects, can use cellulosic material as food, that is 'destroy' it or at any rate reduce its bulk; but it is the fungi that are our chief scavengers as far as cellulosic substances are concerned. Fungi can use cellulose and 'woody' materials as food; in so doing they remove debris which would otherwise accumulate so quickly and in such quantities that we cannot imagine the outcome. When cellulose and *lignin* are present as valuable lumber, their consumption by fungi is economically and sociologically important to us. But let us give the fungi their due: what kind of a world would we have – even in the short time interval of ten years – if all the plants that died remained otherwise unchanged?

Go into the deep, cool forest to see the carbon cycle of life, a cycle in which members of the lowly mould tribe play no small part. Great trees that struggled to gain a foothold in rocky soil have lived their span and died, but even as dead wood they represent a mass of carbon-containing material which can yield food to other organisms, mostly the fungi, in the great quest for energy that is basic to life. In the process of 'digesting' part of the woody materials, fungi bring about a gross dissolution of the wood into fine fragments. Thus the crumbling log, which to the poetic eye symbolizes defeat and decay, is adding humus to the soil and increasing the water-holding capacity of the erstwhile barren rocky ground. This in turn makes possible the life of countless other plants such as mosses and ferns. And equally important is the fact that, as the carbon 'body' of trees is reduced to fragments, the fungi, in their own life processes, change some of the carbon to carbon dioxide gas which passes back into the atmosphere and is reabsorbed by green plants, which are all the while making new food from it. The concept of a carbon cycle sounds involved, and of course it is; however, we can easily appreciate the basic idea that fungi are very important to life as a whole in that they help to keep in circulation the carbon currency on which life on this planet is based.

So, if the moulds, rusts, and their relatives often openly declare themselves man's enemy, they are also at times his friend.

Moreover, they may even be his saviour. Hundreds of kinds of the green moulds (such as the one named *Penicillium*) live their usually unseen lives growing upon all sorts of organic matter. Until recently most of the human race held these in contempt except in so far as the moulds are useful in creating the particular flavours of the better kinds of cheese or are a nuisance because they are capable of destroying stored fruits and vegetables. But a few members of the human race have the knack of finding the sweet uses of adversity. Through the astute observations of Sir Alexander Fleming and the labours of many others, men have learned that not all *Penicillium* moulds are a nuisance or at best 'useless.' Some kinds, like *Penicillium notatum*, produce a chemical substance which has been called *penicillin* that can check the development of bacteria. And because some kinds of bacteria destroy human lives, thousands of persons owe their very lives to the drug produced by a lowly mould. It is all very much a matter of view-point!

As a group, biologists have a fair sense of humility. They have at times and with great profit heeded the biblical injunction 'Go to the ant – consider her ways and be wise.' By studying in the greatest of detail a certain fungus (which is a much less highly complex organism than an ant, at least structurally), biologists have been able to push the frontiers of the science of heredity (genetics) far beyond what could have heen envisioned by the most active imagination fifty years ago. And note again the importance of view-point! A relatively short time ago the Pink Bakery Mould (*Neurospora*) was known to take part in human affairs only, as its name implies, as a mouldy pest that invaded bakeries, where it grew on flour and dough causing discoloration and an undesirable taste. In the early years of the twentieth century no one could have foreseen that this utterly 'useless' organism would teach men some of the most fundamentally important facts about the mechanisms of inheritance. Yet this has come to pass; in a more than thousandfold return for the inconvenience and expense it caused for bakers, the Bakery Mould has been one of the most important tools used by geneticists in their probing into the ultimate details of why like begets like.

A few years ago mushrooms were once again being accorded a large share of print and picture, this time in the role of possible contributors to our knowledge of the working of the mind. Wasson, Heim, and others have been writing large books to tell of the discovery of a cult, which may be as old as man himself, in which certain drug-containing mushrooms are eaten by natives of the Mexican highlands during the course of remarkable rituals. The mushrooms involved cause the person who partakes of them to experience amazing hallucinations that seem far beyond the ordinary realm of human experience. Because of the dreams, some Mexican Indians believe the 'sacred' mushrooms confer upon the participant divinatory vision which is derived, they believe, from contact with supernatural power. This is an exciting story, and one important part of it is the possibility suggested by psychologists and psychiatrists that the drugs that have been isolated from the 'dream mushrooms' and analysed may some day be used to throw light upon the complex processes that go on within the human brain.

The reasons for man's interest in fungi are manifold, and they are as cogent as the reasons for studying any other living organisms. Linnaeus, the great Swedish botanist, in 1752 expressed his views on natural objects in these terms: 'New things are always presented for the delight of his [man's] curiosity, lest the threadbare common objects should make him weary.' Some fungi are common enough, at least in occurrence; almost no fungi are threadbare; and they *can* be a delight for our curiosity even if our appreciation, like that of the three blind men, may be restricted because of how we go about considering them.

Gunnery in
the Fungus World

If copies of Figure 2 were to be distributed world-wide with an offer of a prize for the correct identification of the object represented, there would be relatively few claimants for the prize. In fact, when a similar study of the same object was once exhibited at an international photographic salon, a newspaper report referred to 'a beautiful photograph of some curious objects made of blown glass.' It seems difficult to imagine that anyone could examine a photograph of this object without feeling curious as to its identity, especially if the person were told that it is a living plant.

Lack of acquaintance with the fungus *Pilobolus* is not the result of its being rare or hidden in inaccessible regions of the earth. In most parts of the temperate world, wherever there are large herbivorous animals, some kinds may be found. However, it must be admitted forthwith that the fungus grows upon the excrement of animals, and perhaps it is not to be expected that many people would look on a dung heap for an object of great beauty and an example of one of nature's most remarkable 'guns,' which is operated by a light-sensitive mechanism.

Especially in the temperate parts of Europe and America, the little fungus guns grow up frequently on the excrement of horses. One can scarcely refrain from speculation as to what will become of the fungus in the event that all horses are replaced by gasoline-consuming machines!

But if *Pilobolus* is a plant, where are its leaves, stem, and root? It

2 *Pilobolus* seems to epitomize the glass-blower's art, but is instead a most remarkable fungus found on manure. The clear bulbs explode and shoot the spores – produced in the black masses at the top – on to vegetation. This natural fungus gun does not operate by chance but is aimed by a light-controlled mechanism. Magnified about 40 times

has none of these: all fungi belong to a group of the plant king-dom that botanists call *thallophytes* (or *used* to call, because biolo-gists often change their minds about the most appropriate names for groups of plants and animals), most of which lack the leafy ap-pearance of familiar green plants. The greatest and most extensive part of the body of most fungus plants consists of delicate cob-web-like filaments usually not to be seen without a powerful mag-nifying glass. These branching filaments carry out some of the most important life processes of the fungus. They push out like roots (although they are not true roots) through the soil or rotting wood or whatever the fungus may grow upon. The filaments se-crete chemical substances that can break down or digest such solid materials as dead leaves and old wood. The dissolved breakdown products pass into the fungus filaments and are used as food to provide energy, or as building materials for the production of new living substance that fills the filaments. Increase of living material (it is called *protoplasm*) in the tubes enables them to increase in length. The sum of all these processes of getting food and manu-facturing new protoplasm constitutes vegetative growth; it is highly complex, and many of its details remain a mystery.

The beautiful objects shown in Figure 2 represent a different state in the life of *Pilobolus*, for they are concerned with reproduc-tion. With very few exceptions, fungi (including *Pilobolus*) repro-duce by means of spores which are very tiny bits of the fungus, usually smaller than dust particles. Each spore consists of some sort of hard outer case which helps the spore to remain alive by preventing the delicate contents from drying. Inside the case is a mote of living protoplasm with its all-important nucleus. The lat-ter is about one twelve-thousandth of an inch in diameter; yet it contains submicroscopic granules of special chemical materials which make the fungus what it is, or (as biologists would say) de-termine its heredity.

Gunnery in the Fungus World

Spores in an inert or dormant condition float about in the air and may be carried great distances. Like seeds, they will grow eventually when they land by chance in a place where the temperature and moisture are suitable for growth. The more effectively a fungus can disperse its spores, the greater is the likelihood that it will survive generation after generation, because the greater the dispersal the better will be the chance for a few of the millions of spores to find a warm, moist niche in which to grow.

For the purpose of reproduction, the fungi have developed a remarkable variety of devices for producing and liberating their spores. Probably in no other group of plants with the exception of those that produce seeds have there been developed so many types of adaptations for dispersal. Among the different members of the fungus tribe, certain kinds of spores are shot out from catapults, others are propelled through the air by a jet propulsion mechanism, others are carried about by insects which act as the unwitting agents of dispersal, and still others are splashed by rain out of special minute cups that harness the force of falling raindrops.

The adjustment of these spore-scattering devices to the forces of their environment that operate them is, in numerous instances, so sensitive as to produce a degree of accuracy and efficiency of operation that is truly remarkable.

There is no more shining example of this than the dung-mould *Pilobolus*. The shiny black button-like object atop each plant is a mass of spores contained within a spore-case or *sporangium*. The whole black object becomes a projectile which is shot away by the fungus to a distance of as much as a yard or even more. The spore-case is adhesive and, when it strikes leaves of grass or other foliage, it sticks fast. There it remains until it is eaten by a herbivorous animal. In its passage through the animal's digestive tract, the hard outer part of the spore-case projectile is dissolved; the spores, all unharmed, later germinate readily in the animal's excrement.

13

Thus *Pilobolus* shoots its spores on to vegetation for the first part of their home-leaving and depends on the good offices of a cow or horse for the rest of their journey. The mode of hitch-hiking explains why this fungus is found on the dung heap.

This kind of association is common enough among the fungi; there are many sorts that grow upon the vegetable refuse discarded by herbivorous animals, and most such fungi discharge their spores in such ways as to make certain that they adhere to vegetation that will, at some later time, be eaten by the 'transfer agents' – usually horses and cows. Seldom, however, is a dung fungus (*coprophilous* is the technical and perhaps more polite adjective) able to reach its objective with the remarkable accuracy displayed by *Pilobolus*. Its sticky spore masses are not just shot off in *any* direction but, with eighty to ninety per cent accuracy, are shot in the direction of the greatest light intensity.

Consider first how the shooting is done. Just below the black spore mass is a large (relatively, in the microscopic world) elongated bulb known as the sub-sporangial swelling. As the gun matures, sugar accumulates in the bulb and, because sugar has a great affinity for water and, in the world of living things, draws water to itself by a process known as *osmosis*, the pressure inside the bulb (which may be as great as 5 atmospheres) gradually builds up until it becomes a considerable propulsive force. Meanwhile the wall of the bulb, in a circle just below the black spore mass, grows weaker. At a certain moment, the force inside the bulb causes rupture of the weakened wall, the bulb explodes, and the spore-case is shot away to its destiny.

Stranger still is the fact that the fungus gun is aimed. It is an easy matter to observe that the stalk below the large bulb bends readily in the direction of light. If some plants are gathered while very young and still attached to their natural *substrate* (horse dung), they can be placed in a large glass jar kept closed and there they will continue to grow for several days. If the jar is then placed in such a position that light from a window reaches the fungus only from one side, it will be seen that most if not all the little gun stalks grow towards the light. The biologist tells us they are phototropic.

But mere application of a technical term does not explain anything; what is meant is that light acts as a stimulus which influences the growth of the stalk and its bending towards light. Elongation of the wall of the stalk proceeds more rapidly on one side of it than on the other, and so the stalk bends.

An even more striking observation can be made to convince ourselves not only that the stalks bend towards the light but also that their bending results in a highly accurate aiming of the spore-case projectile. If the jar of growing *Pilobolus* plants is placed in total darkness, or covered on the outside completely with black, light-proof paper, it will be seen that the stalks do not bend at all. Then if a small hole is cut in the paper and a powerful beam of light is played upon the hole it can be observed that the sticky spore-cases, when discharged by the guns, adhere mostly on to the glass jar *just* where the light entered and almost nowhere else. Actual experiments of this nature have shown that eighty per cent or more of the spore-case projectiles may be found adhering to the glass in the lighted area.

How this accuracy is achieved is more interesting still. Not every detail is known but enough has been discovered to give a fairly complete explanation of what happens. First, it is certain that the bulb or sub-sporangial swelling acts as a lens; that is, it focuses or directs the light rays which strike it. This is established partly by photographs such as Figure 2. When the picture was taken, the two bulbs on the extreme right were receiving light from a position corresponding to the upper right corner of the photograph. Upon striking the bulbs, light rays were bent by the curved glassy wall as they would be by a magnifying glass. They were bent as they passed through the bulbs and focused or concentrated upon the far side of the bulbs where a bright circle of light can easily be seen in the photograph.

Two bulbs towards the left were receiving light from the same position but it passed through parts of these bulbs where the curvature was slightly different. The light was again concentrated on the far sides but in a more diffuse manner or over a less restricted area.

This evidence is fairly convincing. However, even better proof of the lens effect of the bulb can be obtained by placing one of the clear bulbs in front of a camera lens and using the fungus bulb as a second lens. This has been done, and remarkably clear enlarged photographs can be taken in which the *Pilobolus* bulb acts as another lens additional to that of the camera.

Biologists have examined carefully the contents of the little fungus lens bulb and have found that, at the base, there is a special accumulation of the living material called protoplasm – in this case really *cytoplasm* because the nucleus is not included. The cytoplasmic accumulation at the lower end of the bulb constitutes a special light-sensitive area for it includes special light-absorbing or light-sensitive pigments. As long as this light-sensitive area is not evenly illuminated, it is acted upon by light in some way that causes uneven stretching of the walls of the fungus stalk below. Exactly how this happens is not fully known.

If diagrams representing the spore-gun are made which trace the paths of light rays as they enter and pass through the lens bulb, it can easily be demonstrated that the sensitive cytoplasm is nearly always unevenly lighted. The stalk is nearly always, therefore, being caused to bend; observation shows that stalks are straight only when grown in the dark. It is possible to conclude, then, that movement of the stalk in the light will always be taking place unless and until the light-sensitive part is completely shaded or possibly very weakly but evenly illuminated. Study of the light-direction diagrams shows that the latter condition is met only when the fungus gun is pointing straight at the incident light. Under these conditions, the light-sensitive part is shaded by the black spore mass directly above it and no stimulus to cause further bending passes to the stalk.

People like to ask *why* something happens. This question is the most difficult for the scientist because there is great danger (especially when dealing with living organisms) of imputing to a phenomenon a purpose which may exist only in his own mind. *Pilobolus* guns *do* discharge their spores and the guns *are* accurately aimed in the direction of the light; but to say for certain why they aim at the light is more difficult than to explain the mechanism. The following considerations attempt to answer the why question; and they could be close to the truth.

In a building lighted only by windows, if one were to fire a bullet from a rifle, the best means for ensuring that the bullet would travel as far as possible would be to aim at a window. Only thus could the bullet encounter the minimum of obstruction. The fungus guns develop in masses of horse dung, many of them towards the inner part of the pile. They may be surrounded by grasses and weeds. By having its spore-guns adapted to bend towards the light, *Pilobolus* probably accomplishes the widest possible dispersal of its spore bullets by shooting them in the direction of light rays.

'Certainly the Elephant,' said one of the blind men, 'even if this one belongs to the world of the Lilliputs, is very like an automatically controlled missile!'

'But,' said another, 'it is also an object having sparkling beauty and symmetry, one of Nature's jewels in its own right.'

The Thread of
Fungus Life

The phrase 'the thread of life' might almost have been invented for the sole use of students of fungi, so apt is it for describing the structure and organization of most moulds and their kin.

In the fungus world, life is contained within a thread; it *is* the thread of fungus life. With the exception of a relatively few of the primitive (at any rate, simple) kinds, the bodies of all fungi whether moulds or mushrooms are composed of microscopic threads. The botanist writes 'the thallus is composed of filaments,' but he is only stating the fact in different language.

The fungus growth shown in Figure 3 is a mere speck, barely large enough to attract attention if it were growing on the surface of jam in a pot. Magnified slightly by a microscope, it appears as a beautiful glistening white mass of branching threads radiating symmetrically from a centre where growth began. Magnified still more, as much as a thousandfold, each thread is seen as a transparent tube, branching and rebranching. In each tube is a rushing torrent of the living contents, the protoplasm, pressing towards the ever expanding tip.

To this tubular microcosm, the botanist gives the euphonious Greek-derived name *mycelium*. Almost everyone has seen it growing as a fuzzy mass on damp bread, as a white or grey scum in a jar of preserved fruit, or as the cobweb-like threads attached to the bottom of a mushroom stalk.

Have you the curiosity to look more closely with a magnifying

3 What might be a button with feathers attached is a mass of spores, growing and developing into the silky filaments which are the vegetative part (mycelium) of *Cyathus*. Magnified about 70 times

glass or, better still, with a microscope? 'Ugh! Only mould,' is the usual reaction.

Only a common mould it is, but in its microscopic structure and organization it is no less complex – or only slightly so – than the cells that make up the laurel leaf or those that make up the brain beneath the laurel-crowned brow!

A pity Figure 6 is a 'still life' and not a cinefilm! If it were the latter, the mycelium would be seen growing, spreading rapidly over the clear jelly from which it is being nourished, glistening as the light catches its transparent tubes and the moving, frothy protoplasm within. As it is, words must suffice to catch the wonder of an unheralded microcosm at work.

As with almost all fungus life as we know it today, the life of mycelium is said to begin with a spore (Figure 4). But it does not *begin*, in the sense of arising anew from what was not alive; it grows from a fungus spore but the spore is alive, itself a tiny mote produced by a fungus of an older generation.

Spores are remarkable in many ways. Consider what must be involved in their capability of remaining alive for months and even, in a few examples where actual tests have been made, for as long as twenty years! Some can resist extreme cold and desiccation and during this time their protoplasm – an extremely complex mixture of substances most of which are what the chemist calls proteins – has become dried to a point where the processes we associate with life would appear to be almost impossible. Yet the potentiality for these processes must somehow remain intact, because when, finally, spores are placed by chance where they can absorb water and where the temperature is moderate, they are still able to grow – at least the hardiest of them. Water is sucked up greedily by the dry protoplasm. Molecules of enzymes, the omnipresent activators of chemical reactions in biological systems, are free to move about in the universal solvent, water. Enzymes can

4 *Mycocalia reticulata*: The elegance of simplicity is shown by this single fungus spore, magnified 10,000 times by the scanning electron microscope. Only 1/100 mm in length, the spore is much too small to be detected by the unaided eye.

attack the substances that were stored away in the spores as food reserves in preparation for just such an hour. The solid food can be converted into sugar and other relatively simple substances. Other enzymes take up a complex job of chemistry in which the sugar molecules are torn apart to yield energy to the now active protoplasm. The renewed activity swells the spore contents, which finally burst out of the restraining (but erstwhile protective) jacket. The restraining spore wall cracks and there emerges into the world of biological competition, fraught with countless hazards, a delicate shiny tube (Figure 5) which will grow into the fungus thread.

Does it matter whether or not the secrets of the spore's trick of lying low for so long or the secrets of what happens when active life is renewed are still mostly secret (for very little of this has been reduced to fact)? Is it of any consequence to men, in any day or age, especially in view of the fact that millions, even billions, of spores are born and grow, far removed from men and their affairs, on the moulding leaves of the forest, on the rotting tree trunk, or in the moist soil beneath?

Perhaps it does not matter. Men ignore much of the living world except when it intrudes upon what they consider to be their own realm. When the spore grows upon the very plants that feed them, men cannot afford to ignore their fungus enemies; but when a spore lands by chance on a bacteria-laden culture and shows itself capable of a mighty contest with germs, Fleming discovers penicillin. When a spore lands upon a precious manuscript, grows into a discolouring mould, and rots the paper, inspiration and the record of history are lost. The answer is a matter of viewpoint.

To study a living mycelium as it appears under a microscope is to be awed by the teeming complex activity and to be obsessed by curiosity. A glance at Figure 6 gives but the faintest impression of living mycelium. The protoplasm rushes along the transparent

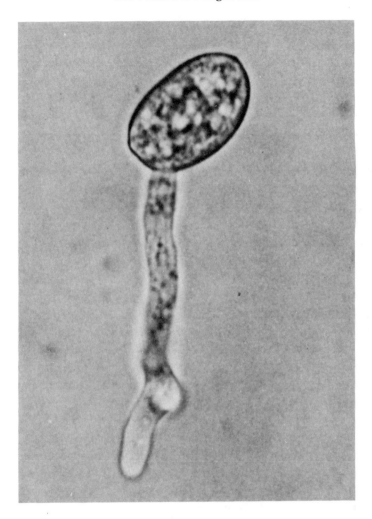

5 *Phillipsia:* The delicate tube that has emerged from this germinating spore will elongate, push its way through soil, and branch repeatedly to form the network of fungus threads (mycelium) that will one day produce a beautiful cup fungus. Magnified 1000 times

tubes or hyphae (each branch of the mycelium is called a *hypha*). Watching a flowing river, one may wonder whence the waters came and where they go, but one might not ask 'what causes the river to flow.' The living protoplasm in a fungus thread is not flowing downhill in obedience to the pull of gravity; it flows up or down, forward or backward, in response to chemical and physical forces of which we know too little.

It is known that protoplasm is a complex mixture of proteins, carbohydrates, oils, mineral substances, and water. It resembles the white of an egg; indeed egg-white is protoplasm, although of a different organism and different in the particular proteins of which it is composed. It is known that the protein complex has the properties of a *colloid* but this name merely classifies it as a substance with large molecular particles having the capacity to dry to the state of hardness, or to absorb water, taking it in among its own particles and holding it so as to form a viscid liquid. As household jelly can flow as a clear liquid when first prepared and set to a semi-solid mound which, when reheated, will flow again, so protoplasm can change from a *gel* to a *sol*, and may move somewhat in doing so. In the fungus threads, protoplasm is not exposed to such extremes of temperature or, as long as it is engaged in active growth, to comparable gain or loss of water.

Protoplasm may also change from gel to sol and back to gel under the influence of electric charges or of an electric current. As a mass of protoplasm moves, changes in electric charges on its surface can be detected; the changes, however, may be the result of movement, not the cause.

One can conclude from watching it that protoplasm may, at times, be pushed bodily from place to place in the fungus threads. In the older parts of the mycelium, watery bubbles often appear in the protoplasm. These are not living protein substance but bubbles of water, containing a variety of chemical salts. These bubbles

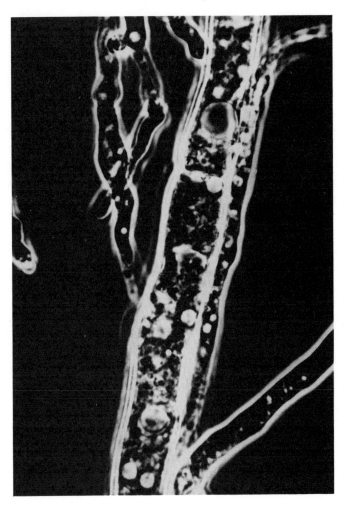

6 *Neurospora:* Not beads and bubbles in a glass tube, but a small piece of fungus thread of the common Bakery Mould. Magnified about 1500 times

(called *vacuoles*) may be seen to enlarge and, as they do, to crowd the living protoplasm out of a particular part of the fungus tube. One might suppose that it is the enlargement of vacuoles which causes protoplasm to move, were it not for the fact that vacuoles are the smallest in the tip of the fungus tube and enlarge in the hindmost and older parts. If the vacuole force explanation were the only one, protoplasm should flow only from the older parts of the fungus tubes to the youngest, most newly formed. But protoplasm does not always behave thus; it can flow towards the tips of the hyphae or away from them, and it may do both sometimes rhythmically – forward for a few seconds, then backward. Moreover, it has been discovered that the protoplasm of certain kinds of fungus or fungus-like organisms contains contractile proteins which are composed of very long molecules which become more viscid or less viscid in response to the presence of certain chemicals. Such a change in the fluid nature of protoplasm may well be another factor contributing to its motion.

Recent study of the structure of protoplasm using one of biology's newer tools, the electron microscope, has given much information about the physical or structural organization of protoplasm. Perhaps soon the clue may be found, but at present it is not easy to put together what is known of the physical and chemical structure of protoplasm to account in a simple way for one of its most basic and important properties, namely self-engendered movement.

The microscope reveals that the protoplasmic river of life is by no means homogeneous. Even the clearest part of it, called cytoplasm, is composed of a granular or frothy material and embedded in this are larger granules that vary in size, shape, and 'shininess,' the latter dependent upon the extent to which light passes through them or is refracted by them (Figure 6). What are all these extremely small bodies (one might well name them 'busy bodies'), many not more than twenty-five-thousandth of an inch wide?

The Thread of Fungus Life

It is easy enough to state that there are two general sorts, a distinction based chiefly upon size but to some extent upon the roles they are known to play; it is much more difficult to state with certainty exactly what function the almost submicroscopic particles may perform. The largest, called *mitochondria*, are commonly rod-shaped as they are seen in living mycelium; some are bodies of irregular shape, and some are spheres. Mitochondria are believed to be essential parts of the chemical mechanism known as respiration in which living cells break down food and obtain chemical energy during the degradation process. The various complex stages of this process are activated by enzymes. Mitochondria seem to be masses of enzymes chiefly involved in the respiration mechanism.

The smaller particles are mostly *ribosomes*. Less is known about them because they are almost too small to be seen even using a microscope that magnifies three thousandfold. Apparently they originate in the nucleus from which they pass out into the cytoplasm. As parts of the cytoplasm they are thought to take an active part in the synthesis of the very stuff of which cytoplasm itself is composed, namely proteins.

With proper lighting and with the experience necessary to distinguish them from vacuoles, *nuclei* can be seen imbedded in cytoplasm. At times they appear quiescent and lie close to the wall of the hypha. At other times they appear to free themselves from the wall and to move rapidly along the hypha, carried with the main stream of rushing cytoplasm, although it is also possible for nuclei to move 'upstream.'

It is usually impossible to discern much detail in a nucleus of a living mycelium, even under very high magnification. Much is known, however, about nuclei, especially those of other organisms such as flowering plants, which may be much larger. Even there, it is necessary to treat them with certain chemicals that prevent shrinkage when stains are applied to give greater distinctness to the delicate structures within.

Study of the microscopic anatomy of nuclei has become one of the most important phases of biological research because nuclei have two all-important functions which, in fact, are closely related: nuclei contain units known as *genes* which determine the hereditary traits of living things, and nuclei govern the ultra-complex chemistry of living cells. With these two concepts in mind, the watcher of the performance of fungus nuclei as seen under the microscope feels excited and curious. What is the meaning of the quiescent period when nuclei remain in the shelter of the slow-moving cytoplasm close to the hyphal wall? What is going on when a nucleus leaves its 'backwater' and moves rapidly along the fungus tubes?

An attempt at a partial answer to the last question brings out another aspect of the significance of the nucleus. When plants or animals reproduce by the sexual process (many can also reproduce asexually), two specialized parts or cells called *gametes* are brought together by one of a great variety of means. In a lily, the pollen grain produces a tube that grows down through the style of the flower to carry into the ovary a male gamete (or sex cell) that unites with a female gamete. In the larger animals, male gametes are motile and they swim until they find female gametes with which to unite.

By whatever means, male and female sex cells eventually fuse into one cell. But this is not all; one of the most important aspects of sexual reproduction, and the one which gives meaning to the process, is the union or fusion of the nuclei that are the most essential parts of male and female gametes. In sexual reproduction, at different times and in different parts according to the kind of organism, a male nucleus must unite with a female nucleus. The two sex nuclei become one, and this one is greatly enriched for the duration of its life because it receives genes representing some of the heredity traits possessed by both parents.

The connection between sex and the movement of nuclei in

fungus tubes is this (though not simply this!). An individual mycelium of most of the larger sorts of fungi, such as mushrooms, contains many nuclei – all of the same sex. Another mycelium has
nuclei all of the opposite sex. The sexes of mushroom mycelia are
not designated as 'male' and 'female,' for no characteristics commonly applied to the sexes can be discerned in either kind. One
must be non-committal and use the symbols + and –, or some
other symbols; the fact remains that the mycelia are sexually complementary. Two different individual mycelia each derived from a
spore of a mushroom must unite in what is certainly a sexual process before new mushroom spores of the next generation can be
produced.

In mushrooms, any part of the mycelium can, in a sense, function as a sex cell or gamete. Any mushroom spore can be considered for simplicity to be either + or – and to produce a mycelium
which is therefore + or –. However, neither kind of mycelium
alone can complete its destiny to produce mushrooms and a new
crop of spores; the two sexes must meet to obey the universal law
of nature.

Such astronomically large numbers of spores are produced that
it is quite probable that a plus spore somewhere will begin to grow
close enough to a minus spore with the result that the two kinds
of mycelia that develop will be able to intermingle. When they do,
the first stage of the sexual process is the fusion of the two complementary kinds of mycelia. It is necessary only for one filament
of a plus mycelium to encounter a minus filament to initiate a process which is stock-in-trade for philosophers and poets.

How this first stage of the bringing together is accomplished is
not known fully in the more highly developed fungi such as
mushrooms. In simpler fungi such as certain of the aquatic
moulds, it is known that chemical substances called *hormones* are
produced. A male filament in water moulds is 'attracted' to or
caused to grow towards a female filament because it responds to

the call (biologists speak of stimulus) of specific chemical substances produced by the female filament. There can be little doubt
that chemical substances are also responsible for the mutual at-
traction between plus and minus filaments of mushroom mycelia.

Anyway, the filaments meet and fuse. Enzymes must pass out
from each to dissolve the solid walls of the mycelial tubes and al-
low the two threads to grow into one. For a crude analogy, think
of two glass tubes being joined by the glass-blower to form one
continuous tube.

When the mycelia have been united by such a fusion, protoplasm is free to move from the tubes of one plant into those of the
other. More important than this is the possibility of mingling of
the two kinds of nuclei, for the union of nuclei is one of the most
important aspects of the sexual process. One major difference between the union of gametes so typical of larger plants and animals
and that which occurs in the mating of the mycelia of mushrooms
is that there are no gametes in the latter example; that is, there are
no specialized cells that serve *solely* for sexual union. Any part of
the mycelium can take on the function of a gamete. There are,
however, many kinds of fungi that do produce specialized gametes.

When sexually complementary mycelia of mushrooms have
anastomosed or fused and the cytoplasm of these has flowed together, it might be expected that the intermingled nuclei of opposite sex would unite at once, as they do in almost all other organisms. They do not, however, but continue to intermingle in a
'social' activity that forms a conspicuous feature of the nuclei in
the mycelia. Plus nuclei wander into and through the minus
mycelium, and minus nuclei into plus mycelium. Such intermin-
gling has been labelled nuclear migration. Many facts about nuclear migration are puzzling. For instance, one would like to know
what forces enable nuclei to travel, for they possess no obvious
mechanism for locomotion. Again, although in certain instances

nuclei from each mycelium pass into the other of the pair, there are numerous instances of plus nuclei migrating only into minus mycelium; here, some forces must direct a one-way traffic – how and why?

Stranger still is the fact that in the mushroom type of fungus, the 'matrimonially minded' nuclei do not unite at once. Instead, they become associated in pairs, one + and one – in each pair. And with decorum they remain as pairs in the mycelium for what may be a long period of growth and development of the fungus threads. Only when the latter have grown into an almost mature mushroom do the nuclear pairs (or their descendants) finally fuse and the nuclear contents of the two parental strains mingle.

It has been suggested that the long period of separate existence of nuclear mates may have a special advantage. If the mycelium containing paired but unfused nuclei should by chance grow towards and fuse with an unmated mycelium (either + or –), the latter could be saved from sterility because nuclei of either appropriate kind would be free to move into the unmated mycelium and establish nuclear pairs there too.

These and many other unsolved problems are among the reasons why the biologist cannot watch the life threads of even the lowliest and simplest of fungi with any feeling of complacency.

If all must be fitted into a pattern, perhaps it could be said that the fungus threads are 'like a tree' in their manner of branching; they are remotely 'like a snake' in their tortuous movement as they grow over decaying leaves, dead wood, or soil; most of all they are among the most amazing of nature's microscopic 'ropes.'

The Mushroom:
A Study in Form and
Function

Writers dealing with natural history frequently point out that familiar aesthetic experiences arising from contact with nature are derived from the beauty of *form*, but rarely do they refer to the beauty of process or *function*. Admittedly, much depends upon one's concept of beauty; to a market gardener, clean, vigorous cabbages in long regular rows may, quite properly, represent beauty, although of a sort not likely to appeal to anyone who happens to dislike cabbage.

Admitted that not everyone can find much beauty in the external features of a Horned Toad or a Gila Monster; yet, recognized or unrecognized, there is beauty inherent in those refinements of the various parts of the animals' bodies that enable them to perform the functions necessary for survival. In the operation of their eyes, for example, there is beauty which is not derived, or not solely, from form.

And, at the risk of labouring the point, there is the possibility that appreciation of nature in terms of the beauty of form only may sometimes result in neglect of a deeper and more lasting source of satisfaction. Egocentric indeed is the person who treasures a wild rose only because of what he or she may deem beauty of form. For its beauty, the rose has, most certainly, a place among beatitudes, but the rose has also its own place as a living organism with intricate adaptations of structure and function that should command respect; the fundamental biological problems of a rose as a fellow being of man are different from man's mainly in number and magnitude rather than in kind.

The mushrooms shown in Figure 7 have, to be sure, certain elements of beauty of form, although not every photographer is able to appreciate and capture them. Beauty indeed there is in the solid, terete columns that appear (if dimensions are forgotten) capable of supporting masses of masonry. The 'heavy' columns are the fungus stems (or *stipes*). Their marble-like appearance is deceptive, for they are formed of myriads of delicate interwoven fungus threads that also compose the rest of the mushroom. Beauty there is too in the clean straight lines that radiate symmetrically from the stem, on the underside of the mushroom cap. These are the *lamellae* or spore-producing plates of the fungus. However, except in the view seen by an artist-photographer, most mushrooms have few features with aesthetic appeal; they might even be described by some as massive, plain, or blocky. Some mushrooms are beautiful because of their colour, but very many of them are pale grey or drab.

When it comes to function and processes and the intricacies of structure related to function, he who cannot find beauty in the drabbest mushroom needs to look again and look also to his own ability to perceive. For a mushroom is a marvel of organized economy and of precise adjustment when examined from the point of view of the function it performs, which is the production of millions of spores for the propagation of the fungus and the efficient delivery of spores to the winds and gentler air currents for the spread of the fungus over the earth.

Mushrooms represent the culmination of a long period of activity of the mycelium from which they spring. In the soil among the grass roots of a meadow or in the mat of dead and dying leaves beneath trees, mycelium spreads out and adds to its mass day after day, perhaps for months. By the chemical machinery within the threads, carbon-containing and nitrogen-containing materials –

7 Stalks, caps, and spore-producing plates (lamellae) of the mushroom *Clitocybe* could be a modern sculpture in marble or alabaster. About three-quarters natural size

the remains of wood, leaves, and fruits – are converted into other kinds of carbon and nitrogen materials to make up the transparent walls of millions more fungus threads and the ever flowing cytoplasm and nuclei within the threads. Some of the incoming materials are converted into sugar, fat, and a substance known as *glycogen*. Any surplus remains in the fungus threads as a reserve of food materials, needed literally as well as figuratively for a rainy day.

Mycelial mates, plus and minus, meet in the ground and nuclei associate in pairs in preparation for the final acts of reproduction to be carried out in the mushrooms-to-be. There comes a day when the mycelium has done its work, earthy but essential. At several points throughout the network small knobs or tangles of threads arise, each a potential mushroom. What determines how long the mycelium must grow before formation of mushrooms can begin cannot be stated in general terms; the time varies according to the kind or species, and with circumstance. For one thing, sufficient protoplasmic material and food reserves must be built up. For another, the temperature and moisture conditions must be propitious; in dry weather the mycelial maze may remain inactive for many weeks, and it is only when the rains come that mycelium can put forth its supreme effort of producing mushrooms.

In wet weather, the tiny mycelial knots or tangles develop rapidly into 'buttons,' little masses of fungus threads closely interwoven to form bodies of considerable solidity and so organized and so co-ordinated in activity that already some threads are destined to produce the cap, some the spore-plates, and some the stem.

Then, rapidly, the cap expands and there are formed within it cavities in which the all-important spore-plates develop. From the roofs of the microscopic cavities, like stalactites in a cavern, certain threads grow downward and begin to form the spore-plates.

The Mushroom

Only the most rudimentary facts are known about the formation of even the commonest mushroom. Mycologists (students of fungi) have described in great detail the various stages in the development of mushroom structures; painstakingly they have given names to all the microscopic parts of mushrooms; most assiduously of all, they have spent countless hours giving names to the thousands of kinds of mushrooms and devising systems whereby the kinds can be identified by the colour of the spores and the external and internal structure of the spore-plates and the stem. Books dealing with this sort of knowledge are legion, yet too seldom have questions been raised as to the internal control mechanisms that result in a mushroom being capable of developing as it does.

To pose a few questions in simple form, consider first the initiation of mushroom development. Exactly what changes are involved that lead from the growing or vegetative phase (of mycelial growth) to the reproductive phase? For weeks or longer, the mycelium has been spreading throughout its microscopic realm, probing every corner for new food, synthesizing food reserves, branching and anastomosing, every anastomosis making new bridges for the free intercourse of nuclei and cytoplasm. Then suddenly all this is changed; activity is no longer 'directed' towards further vegetative growth but towards the production of mushrooms. Yet mushrooms are composed only of intertwined hyphae. One day, hyphae are growing towards sources of more food and *away* from one another, partly as a result of avoidance of their own chemical by-products, and so are pushing radially ever farther from their point of origin; the next day some hyphae are growing in a highly co-ordinated manner *towards* one another and they intertwine with one another to build up a solid mass which the mycologist calls the *stipe* (stem).

Regard for a mushroom as a living entity gives us courage to hazard some guesses as to what may be happening, but it must be admitted that there is deplorably little proof. Recall what can be seen through a microscope by anyone, namely that the vast microscopic complex of fungus threads forms a system in which there is possible a free flow of materials from any one part to any other. Although each tube or hypha of the mushroom mycelium is divided by partitions or cross walls into units (or cells), each partition between cells is perforated by a hole through which protoplasm flows freely along the fungus tubes. Hyphae branch in all directions and, because the branches can fuse or anastomose, there is interconnection between most, if not all, of the main branches. The mycelial network that pervades an area of several square yards of forest floor is truly as much an intimately connected system as is a great railway or a telephone system that covers a continent.

When (and we do not know *why*) the activity of the mycelium has become centred in the production of a mushroom button at a particular point in the network, it is not necessary for millions of new fungus threads to be produced out of decayed leaves and wood. The protoplasm can move to the activity centre – the growing mushroom. Returning protoplasm 'called in' from all the outlying parts can fill the rapidly developing hyphae that are weaving themselves, following a fixed pattern, into a mushroom. Whether all this really happens or not, it is inconceivable that a mushroom can grow as quickly as it does from a button to a large intricate structure entirely by the production of new material at a given point. Under the optimum conditions of environment and with maximum feeding, the mycelium of ordinary kinds of mushrooms seldom grows more than one-sixth of an inch a day; mushrooms themselves, however, appear like magic – almost overnight.

Or do they? Certainly they develop so rapidly that the expression 'pop up like mushrooms' has become a standard one. Careful observation, however, shows that several days may be required for the merest button to grow into a mature mushroom. In its early stages of development, the little mushroom may easily be over-looked as it nestles in the grass or lies partly overlain by leaves. It is only in the later stages of growth that mushrooms appear to 'pop' up, and that because the stalk may elongate quite rapidly. Besides, mushrooms are not always able to develop continuously in the course of a few days; their rapid growth may be arrested by a dry spell and may not be resumed until rain comes again.

As the spore-plates mature, special hyphae on their surfaces develop rapidly into microscopic club-like bodies called basidia. Ultimately each club produces four spores at its apex, each spore attached to an extremely delicate prong protruding from the basidium (Figure 9).

Meanwhile important events have been taking place inside the basidium itself, events that represent the culmination of the earlier fusion of plus and minus mycelia and the pairing of nuclear mates. Each young basidium receives one pair of sexually compatible nuclei. Then at last, after perhaps weeks of association as a pair of individuals, the two nuclei that ultimately find themselves in each basidium truly unite. The event that usually happens soon after two sex cells unite in most kinds of plants and animals, namely the mingling of the heritable materials derived from both parents, happens in the basidia of mushrooms only after a remarkably long 'courtship.'

The single nucleus produced by fusion of the nuclear mates then undergoes division to form more nuclei that are needed for the spores. These divisions comprise one of the most remarkable and exact processes in the living world. They result in a shuffling and redistribution of the heritable materials called genes, as a consequence of which the newly formed nuclei may contain different combinations of genes. After this, one nucleus moves into each spore. Depending upon the genes present in the individual nuclei, a spore will grow in due course into a plus mycelium or a minus. Because of this gene shuffling also, mycelia may differ in other respects such as colour, rate of growth, food requirements, and many more.

In the Field Mushroom or the common Cultivated Mushroom the spore-plates at maturity are still covered by a delicate membrane. This is torn as the mushroom cap expands (Figure 8) and in some kinds of mushrooms remains as a ring of soft tissue adhering to the stem just below the cap. The spore-plates are then exposed to the outside air and are free to discharge their spores. Although the spores of many kinds of fungi have no special way of becoming detached from the parent body, mushroom spores are *shot* away from their basidia (see Figure 9).

It is not difficult to observe this extraordinary process. A spore-plate is detached carefully from a full-grown mushroom and placed in a shallow glass vessel so that the fungus tissue does not die by drying and a microscope is brought to bear upon the surface of the spore-plate; basidia, each with four spores, can then be seen. A single basidium is kept under observation and, if all goes well, one spore is seen to fly off suddenly from its attachment, then a second; finally all four are gone. They did not fall off, for it can be seen that they fly up towards the observer if the spore-bearing basidia face upwards. Quite often the basidium can be seen to shake as though by recoil as each spore is shot away. Years of painstaking investigation have been devoted to observations of the violent spore discharge of mushrooms, and many pages of print report the observations; still the mechanics of the process remain somewhat uncertain.

Mushroom spore-plates hang vertically downwards from the underside of the mushroom's cap. Spores are shot off from their basidia out into the still air between two closely placed adjacent plates, and fall under the influence of gravity until they are free of the plates, free to be carried away by air currents. Anyone venturesome enough to go into the woods on a warm autumn evening can see spores being freed from the underside of a large fresh

8 *Pholiota:* Four gnomes, fur-bundled against the winter cold, are mushrooms that can be found in summer in almost any woodlot. About three-quarters natural size

mushroom. Under the beam of a strong flashlight against the velvet blackness of the forest at night, spores can be seen as a fine mist streaming away from beneath the mushroom parasol. Evidence may be more conveniently obtained by cutting the cap from a fresh mushroom and placing it, spore-plate side down, upon a piece of paper. Before the mushroom cap dries, thousands of basidia will discharge their spores, which then settle on to the paper and make what mycologists call a spore print. The spores pile up on the paper in lines corresponding to the spaces between the thin spore-plates.

The form and the manner of development of a mushroom assume new significance when they are thought of in relation to the process of spore production and liberation. Consider for instance what would happen if the spore-plates did not hang precisely vertically down. They develop very close to one another, often less than one-sixteenth of an inch apart. After spores are shot free of their basidia, they fall in the still air between the spore-plates and eventually reach open air below. If the mushroom cap were tilted away from a vertical position, even slightly, falling spores would lodge on adjacent spore-plates. However, few spores (if any) are wasted in this way. During the early stages of its growth, the stem responds to the influence of gravity. As the stem elongates, it adjusts itself constantly away from the downward pull of gravity. The cap too must react to gravity, for its tissues are spread out horizontally. The spore-plates themselves respond positively to gravity. Thus, in the mature mushroom, the stalk bears the cap well above the ground, allowing spores to be caught by breezes, the cap expands horizontally, and the dozens of thin vertically oriented spore-plates are free to loose their spores.

The young stems of many kinds of mushroom are sensitive to yet another factor of the environment, namely light. If a group of mushrooms is observed that began to develop in the confined

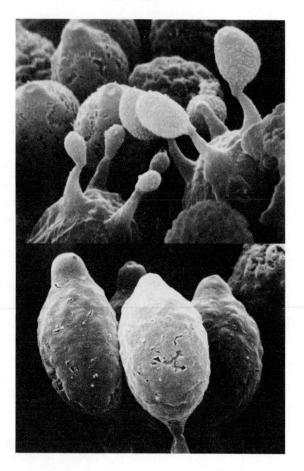

9 *Lacrymaria velutina:* Basidia, the producers and projectors of spores of a small parasol-like mushroom, as revealed by the scanning electron microscope. At the top are shown two basidia: *left,* the four prongs, each bearing young spores; *right,* four partially mature spores, magnified about 4000 times. At the bottom are shown, at a magnification of 8000 times, four mature spores ready to be shot off from their basidium.

space of a dark recess in a hollow log, it can be seen that the stems grow towards an opening. The caps are developed when the stems are able to carry them out into the open by growing towards the light and the stems may often be seen to have undergone much twisting during the process. And what seems certain from such observation has, in fact, been verified by careful experiment. Although most mycelium in the vegetative phase of growth does not grow towards light (and may even grow away from it), the young stem (part of the reproductive phase), although it too is composed of mycelium, is strong in its positive growth reaction to light. Surely there is material here for thought, and plenty for further investigation.

Even the fact that mushroom spores are developed on thin plates or lamellae deserves consideration, for spores are not always developed on such plates even among many large fleshy basidium-bearing fungi to which mushrooms are most closely related. Many mycologists take the view, which seems reasonable and capable of support, that the lamellae of a mushroom represent the maximum surface for spore production. Certainly the total surface area, counting the two sides of each thin spore-plate, is many times that of a plane surface such as that of the smooth upper side of the mushroom cap. There are other basidium-bearing fungi in which the spore-producing surface seems to have been made greater by the development of hills and valleys, or of low ridges. In still others, tooth-like projections (Figure 10) provide a large surface. Many mycologists interpret these stages of area increase as indicating the consequences of evolution and maintain that the maximum spore-bearing surface represented by the spore-plates of common mushrooms has strong survival value for such mushrooms. They would argue also that, in this respect, plate-bearing mushrooms are more highly evolved than their 'flat-faced' relatives.

The Mushroom

Even the hundreds of students of mushroom lore who concern themselves mainly with finding the correct Latin name for their collections must admit that countless processes go on inside mushrooms about which very little is known. What little is known of function constitutes beauty – the beauty of co-ordination, precision, adaptation.

As for the blind men – perhaps mushrooms remain for them mainly tasty morsels, and not much 'like' anything else in the wide world!

The Coral Fungus

There is a tradition among students of the fungi, and it is also a recorded fact, that the experience of finding for the first time the exquisite Coral Fungus (Figure 10) in the cool, moist woodlands of Sweden aroused the interest of Elias Fries to such a degree that he became one of the founders of mycological science. The first sight of any beautiful natural object can arouse emotions and thoughts never quite duplicated later when one has become more familiar with it and has realized that it is probably not so rare as anything so peerless must surely seem to be. Even in a photograph (Figure 10), *Hericium coralloides* can produce the same effect upon the beholder today as it did upon the great Fries over a century ago, though the same ultimate result can hardly be expected.

And, because the recollection of a first impression usually brings to the mind's eye a host of other memories inextricably associated with it, one is easily transported to the time, the place, and even the attendant circumstances of that first memory imprint of 'a day from many singled out.' One sees again the autumn forest floor, dappled with golden sunshine, bedaubed with the pigments of early fallen leaves where, only here and there, are dimensions and perspective established by flaming crimson dots of maturing bunchberries or luminous pads of chromegreen moss. One is aware of that most pungent attar which is the bequest and benediction of dead and dying leaves. One hears again the rustling of dry leaves as a grouse struts cautiously away, and then shatters

10 *Hericium coralloides:* Hoar frost in summer is suggested by the delicate branches of the Coral Fungus. About one-half natural size

the near silence in an explosion of whirring flight. One recalls a red-brown crumbling log, on one end of it a group of fragile toad-stools, silvery grey and glistening; and on the other end – a wonder! – an object of transcendent beauty, defying description, defying credulity even! Someone, a long time ago, must have thought that it resembles a white coral, for that is the meaning of the name it bears – *coralloides*, 'like a coral.' Perhaps it does resemble a coral; it is composed of an infinity of delicate pendant branches which rebranch into more delicate branches until they become so tenuous that they seem almost to disappear. But no true coral gleams with the pearly translucence of the living coral before our eyes. What so exquisitely delicate *can* it be, other than the tracery of a crystalline frost painting on a window pane? For one mycologist, *Hericium coralloides* suggests a 'diminutive water-fall frozen in full fall but made of milk, not water.' And, although little would be learned from the exercise, it might be interesting to know what a person might think about the Coral Fungus who had no simple basis, at least not at first, for answering the question 'is it animal, vegetable or mineral?' – as it is often put in guessing games.

The living coral of Figure 10 is a plant – a fungus, related to that assemblage of large fleshy fungi that includes the mushrooms. In its most basic respects it is like a mushroom because its spores are developed on the ends of prongs that protrude from the top ends of microscopic club-shaped bodies. Also, like a mushroom, the Coral Fungus begins its life as an air-borne spore which may have journeyed far and ended by landing on the mouldering log. Then, at first from its own resources, the spore sends out a minute glistening tube which extends hourly until, probing the soft decaying wood, it finds moisture and food. The tube grows and becomes the branching, spreading filaments we call mycelium in which takes place the business of disentangling the molecular complexes of the wood, making use of what is available as food and transforming that into more fungus substance by incredibly complex alchemy.

In time, when the requirements of several niceties of temperature, moisture, and chemical changes have been met and the time for reproduction is at hand, mycelial strands of the Coral Fungus – like those of the mushroom – become organized around a focal point which eventually assumes the form of the mature coral.

In the earliest stages of its development, when the Coral Fungus is still only a small ball of interwoven mycelial cords, it resembles the comparable stage (in the matter of size) of a mushroom, which we called the 'button' stage. No one would be able, by visual examination, to predict whether such a button would develop into a Coral Fungus or into a mushroom. Moreover, the superficial resemblance does not end there; each young fruit body, the button, is constructed of mycelium or fungus threads which are so similar in each kind of button that it is doubtful that even experts could distinguish the two kinds. Despite these apparent similarities, however, there are fundmental invisible differences between the mycelium whose threads form the warp and woof of a Coral Fungus and those which form a mushroom. The basic construction elements, the mycelial threads, of the two fungi contain different kinds of nuclei, not in the matter of the appearance in general or the gross structure of the nuclei but in the nature of the genes which they contain – genes that control heredity and determine form. The particular genes present in the nuclei of the Coral Fungus, basically, govern the chemical activities of its mycelium and determine the pathways and patterns of physical development. Although there is abundant conclusive evidence that that is true, there is little evidence concerning the exact nature of the chemical reactions involved and concerning the means whereby mycelial growth is directed and modified. It may therefore be wiser, and certainly simpler, to consider a few more details about the end product, the chaste plant coral itself.

The Coral Fungus

As we have discovered, the Coral Fungus reveals its relationship to and affiliation with other groups of fungi in a single microscopic feature, namely the manner of development of its spores on basidia. A mushroom's spores are also developed on and liberated from basidia of almost identical form; but whereas the basidia of a mushroom cover the surface of the thin plates (called gills) that hang vertically on the underside of the expansive mushroom cap, the spore-producing basidia of the Coral Fungus cover the outer surface of its delicate pendant branchlets. In mycological writing, the branchlets are usually referred to as 'teeth,' a description which seems discordant when applied to the milky icicles of the Coral Fungus. Be that as it may, the pendant branches do bear the spore-producing basidia from which, presumably, spores can be shot off into the air as conveniently as from the plates or gills (another term that refers to animals!) of a mushroom.

Presumably the branchlets of the Coral Fungus grow downward in response to the effect of gravity, though it appears that no one has proved this by actual experiment. It is an amusing thought to realize that if, by some mishap, the Coral Fungus pushed its most delicate branches upwards instead of down, it would be classified in a different family of fungi – the Clavariaceae – in which the ultimate branches are upright, not down-pointing as they are in the family Hydnaceae to which the Coral Fungus is considered to belong. It would be interesting also to know what brings about such a difference in a growth response in the two families of fungi.

Most fungus books have little to report about the Coral Fungus beyond giving descriptions of form, dimensions, size of spores, and other similar details necessary for identification. Some of them, however, do report that the Coral Fungus is edible. When confronted with starvation, humans rate the edibility of something very high among its virtues; under any less critical circumstances, the edibility of the Coral Fungus must surely appear as a feature unworthy of being extolled. Who would grind a carved

Chinese jade into powder because it might be made into green paint?

Happily, the dicta of the blind men – or any mere words indeed – cannot detract from the absolute loveliness of *Hericium coralloides*. When in any one year it has disappeared, through decay or under the blanket of the first snow of autumn, its imprint of beauty remains on our memories – a joy forever.

The Mystery of
the Water-Guns

Probably not one of the three analysts of our old fable would have been able, by his own devices, to assist us in ascertaining the *like-ness* of the fungi represented in the frontispiece of this book for the simple reason that even to find them requires sharp eyesight, which no one of them possessed. But had a person with normal vision found the fungi, presented them to the blind men and begged for a pronouncement, there might well have been heard such descriptions as these:

'Like a *fairy goblet*, by its size and shape.'

'Like a Lilliputian *bell*, upside down.'

'Like the *nest* of the smallest ever of all *birds*, for even the bird's eggs lie within.

And, feeling the eggs within the nest, one of the men might even have added: 'It is truly like a plant; are not the tiny eggs like *seeds* of mustard?'

There would have been some truth in what each man reported. Terms such as theirs have been used for a very long time to name and describe what are today most often called Bird's Nest Fungi. Since 1601, in fact, these curious little fungus cups have been known to botanists. However, the significance of the vase-like or cup-like shape of these small natural curios and especially the true nature of their seed-like contents have not been fully understood until very recently, three and a half centuries after Bird's Nest Fungi were first described.

The reason that Bird's Nest Fungi remained so long shrouded in mystery is not that they are so rare anywhere or that they may be found only in inaccessible places in tropical jungles; on the contrary, two kinds are often found in temperate regions in vegetable gardens or around old compost piles and even on old damp boards in shaded locations. The little cups are seldom more than a quarter-inch in diameter and one needs to look carefully.

No, indeed, scarcity of material for study was not the barrier to the solution of the mystery; it was scarcity of perception that kept the door locked or, as modern journalese would have it, prevented a 'breakthrough.' Although it is commonly believed that scientific discoveries are the result of some sudden and strange enlightenment, scientists themselves recognize that what usually happens is quite otherwise. Patiently and piece by piece a jigsaw puzzle must be put together before there can appear a complete and meaningful picture. One investigator puts a piece or two in place, and another adds a few more pieces which seem to fit; but very often the picture cannot be completed until some one key piece is found. Even when it is found, it may not appear at first to fit into the whole. This again may be a matter of viewpoint; the piece may appear very different when seen from one position than it does when seen from another – possibly it cannot be seen to fit merely because it is being viewed upside down in relation to the whole. But when the last piece of the puzzle is properly in place, the entire picture becomes complete and meaningful and a discovery may result. Note the word 'may.' Even when all the evidence is before a scientist, a new idea or concept may not come forth unless the observer possesses the ability, by virtue of training and a natural or cultivated capacity for deduction, to appreciate the significance of the evidence.

The Mystery of the Water-Guns

But returning to the Bird's Nest Fungus, a bit more about the mystery: why was there a mystery and why were botanists so interested in solving it? In the middle eighteenth century and beyond, indeed well into the middle of the nineteenth, biologists were trying to fit animals and plants into the various schemes of classification which had been devised. Among plants, one of the major divisions of classification was the result of the endeavour to distinguish between those which produce seeds and those which do not. Fungi do not reproduce by means of seeds, but by spores which are much more minute (microscopic in fact) and simpler than seeds. Bird's Nest Fungi smell like mushrooms, feel like some other kinds of fungi, and, when very young, resemble tiny puff-balls; for these and other reasons they were classified as fungi. But fungi were not known to produce visible reproductive bodies that look much like seeds; and out of this enigma arose a heated and long-protracted controversy as to whether the 'eggs' in the Bird's Nest Fungi are seeds or are not seeds.

By the early nineteenth century, the microscope had been sufficiently refined to enable those who used it to study very thin slices of the supposed seeds of Bird's Nest Fungi to see that no resemblance to a seed could be found. Rather, the little eggs were clearly seen as capsules containing myriads of microscopic spores resembling those of many other kinds of true fungi; moreover, when the spores had been germinated, it was observed that they gave rise to filamentous growths very similar to those of other fungi whose spores had been seen to germinate. So the Bird's Nests came to be recognized as authentic fungi, despite the superficial resemblance of their little spore cases to seeds.

But the solution of that problem gave rise to another which proved to be an even more difficult one than the first. Microscopists who peered long and carefully at the seed cases also learned much about the structure of the entire fungus nest and its eggs; what they learned was intriguing but left them more mystified than before. In the first place, it could be seen that the spores were securely locked inside the very thick strong wall of the capsule within which the spores had been developed. Spores of fungi serve not only for reproduction but also for distribution of the fungi that produce them, often very widely over the earth. But how could the spores of the Bird's Nest Fungi serve for distribution when they are securely locked within a capsule which seemed to have no visible escape hatch? Not only that, the entire spore-containing capsule itself (at least in some kinds of these fungi) was observed to be apparently fastened by a kind of stout cord or stem to the inner wall of the cup or nest. Seen under the low power of a microscope, the capsules and their attachment cords bear a crude resemblance to a heavy button, with its stem made of thread, firmly attached to an overcoat. More careful study at higher magnification reveals that the fungus attachment stem possesses a complex internal structure. The most essential part of the internal detail is a long, astonishingly strong, microscopic cable made up of fungus filaments. If one dissects the fungus apparatus carefully, the cable, which was neatly folded inside the stem, can be pulled out to a length of as much as four inches. That is long considering that the whole cup is but a quarter-inch in diameter and a spore capsule only a bit over one-sixteenth of an inch. Naturally, inquisitive botanists wondered mightily about the possible function (people often say 'use') of such a piece of machinery.

One botanist, playing the role of one of our blind men quite unaware, announced that the stem containing the long coiled-up cable must be some kind of spring mechanism by means of which the spore-case could be shot away from the little fungus cup – obviously a wild guess in view of the fact that he did not see any capsules being shot off. Another botanist, who probably thought the opinion of the first one was ludicrous, declared that what must happen is that insects, interested in the Bird's Nests as food, get their legs or other body parts entangled in the long cable and so carry off the spore-cases to distant places.

Other guesses too were made; but in time pieces of evidence began to accumulate so that, about 1940, it was possible to add them all up to complete an astonishing story which has since been verified by careful observation and experimentation.

The little symmetrical vases (see frontispiece) when fully mature and open, are what have been called 'splash cups' which make use of the energy of large falling raindrops for the ejection of the seed-like spore capsules to a distance commonly of 3 to 4 feet and occasionally as much as 7 feet. A heavy raindrop is falling fast and, because of its weight, it represents a considerable force in the world of the small things. The Bird's Nest Fungus cups are just the right size to receive large raindrops; and, if a drop lands squarely in the centre of a cup, spore-cases are spattered out of the cup. But it is by no means just a spattering. When water falls into the cup, a splash is created during which water drives up along the sides of the cup, tears at the cord by which the spore-case is attached, and flies away from the cup carrying the spore-case, the latter with part of its cord attached. One end of the cord is very sticky and when this end happens to touch a leaf or branch nearby, the cord adheres to it. The spore-case has weight and, because it is travelling quite fast, it tends to continue in its path of flight and the long wound-up cable is pulled out. When the spore-case has reached the end of its rope it is jerked back, of course, and this jerk causes it to be wound around the attachment point. The latter happening may be compared roughly to the action of the bolas used in South America to catch running animals: a stone or some other heavy object is tied to each end of a rope about six feet long; when the bolas is hurled at the animal's legs, the rope strikes the legs and the heavy stones or balls tend to wind around the legs thus throwing the animal to the ground.

The spore-case attached by its long adhesive cord remains fastened to a leaf or some other plant part until the latter is eaten by some herbivorous animal. The fungus spore-cases are of course eaten by the animal along with the leaves and pass into the animal's digestive tract. There, the outer wall of the spore-case softens and breaks down and the spores are freed and ready to germinate in the animal's feces wherever these are dropped. Ultimately, the spores germinate and develop into mycelium from which will be formed a new crop of Bird's Nest Fungi. Thus, reproduction and dissemination of this organism have been accomplished. Small wonder that the particular Bird's Nest Fungus in the frontispiece is occasionally found on old manure or on garden soil to which manure has been added. The first part of the spore distribution process depends upon the splash action of raindrops, which accomplishes a very local distribution; for the second part of its ride, the spore-case is spread even more widely through the unwitting assistance of a horse, a placid cow, or a wild animal.

All this is not a haphazard business – far from it. Study and experiment have shown that the fungus cup, in size, is just what is needed to make the best use of the energy of raindrops. Moreover, the angle of the sides of the cup is just that which is needed to create the best splash in the water-'guns,' that is, to accomplish the farthest lateral throw of the little spore capsules.

This story can easily be verified, at least part of it, by anyone who has been fortunate enough to find some Bird's Nest Fungi and is willing to stand outdoors in a pelting rainstorm to watch for what can happen. When a raindrop lands in a cup, there is an audible 'plop'; after a few more direct hits the cup is empty. Where are the spore-cases now? If one searches diligently on leaves, stems, and grasses all around the spot and even up to four or more feet from the cup, spore-cases will surely be found with their microscopic silken cords wrapped around a twig or entangled among the hairs on the surface of a leaf.

The Mystery of the Water-Guns

Although the 'nests' of all Bird's Nest Fungi contain spore-cases (unless the nests have already been emptied by raindrops), in some kinds (species) the spore-cases do not possess special attachment cords. In such species, the spore-cases themselves are adhesive all over and, when splashed by rain on to nearby vegetation, they can easily become attached to anything with which they make a flat or broadside contact.

This account of rain-operated spore dispersal in the Bird's Nest Fungi is but a brief and simple version of what is known and what can be read at length and in detail elsewhere. Even so, it may be sufficient to engender some thoughts in the philosophically minded reader. One cannot escape a feeling of profound respect for those forces of nature which have operated to bring about so neat and so efficient a correlation between the urgent and vital process of the dispersal of reproductive units (in our example, spores) and a naturally occurring and readily available source of energy for accomplishing the process. One is delighted by the beauty inherent not so much in the form or colour of Bird's Nest Fungi as in the means whereby they are able to compete successfully with other living things in the great contest for continuing survival.

Biologists do not usually rest contented with philosophical considerations, however. They are more inclined, on becoming aware of a previously unrecognized structure or process, to examine other organisms in the light of the new viewpoint or new discovery. 'How widely spread is the occurrence of that structure or this phenomenon among organisms?' they ask, and 'might the occurrence of this phenomenon in different groups of organisms offer any clues as to possible evolution or relationships between different groups of organisms?'

And so it was for the story of the Bird's Nest Fungi. When their little cups had been recognized as splash 'guns' that bring about the scattering of spores through the agency of raindrops, botanists began to re-examine other cup-like structures in the plant world. Seen from a fresh viewpoint, small plant cups which had long been so familiar that they no longer commanded much attention were considered worth a second look.

A little bracket-like fungus that grows on dead twigs had always been a curio because, in addition to producing its shelf or bracket on the underside of which spores are developed as in all the members of that fungus group, this particular one also produces peculiar tiny cups. From one lip of the cup, the bracket part eventually grows out, as every student of fungi knew. But why the cup – was it of any 'use' to the fungus? Why, especially, because the cup bears such a close resemblance to a Bird's Nest Fungus that it had been thought to be one of them by more than one fungus specialist. The cups of the bracket fungus, however, seemed always to be empty when found. But one botanist – who was completely convinced of the importance of the splash-dispersal concept – persevered by collecting the cups of the bracket fungus almost every day of the year when they could be found. It turned out that the cups are not empty if you look at them in the early autumn; they are filled with microscopic rod-shaped spores. A little careful study and some simple experiments were all that was needed to add the bracket fungus to the list of splash guns. It is, of course, a different fungus and one rather distantly related to Bird's Nest Fungi.

What about splash guns among green plants which are very different from fungi? A small, primitive green plant called liverwort or *Marchantia* that grows in boggy places, it was recalled, produces minute green cups inside of which are borne tiny lens-shaped masses of green tissue which are known to function reproductively. Although the *Marchantia* cups are smaller than those of Bird's Nest Fungi, the angle of the cup sides is the same in the two kinds of organisms and the cups have the same organization in general. Do the cups of *Marchantia* operate in essentially the same manner

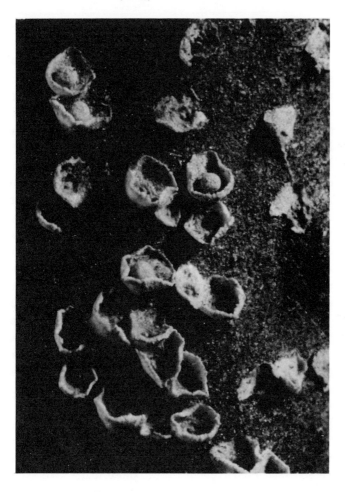

11 *Laeticorticium minnsiae:* A dainty flesh-pink fungus, about pinhead size, growing on the upper side of a hemlock twig. Each cup contains a single ball of fungus tissue which is thrown out from the minuscule splash gun by raindrops to start a new growth of fungus. Magnified about eight times

as in Bird's Nest Fungi? Not much examination or experiment was needed to establish the liverwort in the category of splash guns. The green lens-like plant bits are easily splashed by rain out of the cups in which they are held.

Botanists also recalled that the male sex cells or sperms of mosses such as *Polytrichum* are developed in the centre of a daisy-like rosette of special stiff moss leaves. Another possible splash gun? It is of the right size and shape, though the sides may be a bit too flat. However, again a bit of study and some 'splashing experiments' easily demonstrated that the male plant rosettes of that particular moss are splash guns. In this example, the splash concept seems to throw some new light upon an old problem pertaining to the mosses. Male gametes must swim in water in order to reach the moss egg-cells during the fertilization process. But because *Polytrichum* plants are either male or female and the male plants often grow at some distance from the female, there seemed to be a problem about transport, though there was the possibility of a male gamete (which is microscopic in size) swimming down on the male plant to the ground, across the ground for a distance of at least a yard (occasionally, anyway), and up on the stalk of a female moss plant to reach the female gamete or egg cell. Experiments established that such a circuitous and hazardous journey may be unnecessary: male gametes can be *splashed* from the moss water-gun as far as four feet; and so they can literally rain down upon the female plants and find themselves very close to egg cells in short order.

The so-called higher green plants, those that produce flowers and seeds, are famous for the variety of devices they possess for the scattering of their seeds: those of the dandelion ride far upon the breezes by means of exquisitely fashioned parachutes; those of the cocklebur and other similar plants hitch a ride by becoming attached (by means of hooks or barbs) to any fur-bearing animal that happens to pass; some mistletoes produce sticky seeds that cling to the feet of birds as a means of transport; and many kinds of flowering plants produce their seeds in capsular fruits of many sorts from which the seeds, at the proper time, are propelled by a variety of catapult and explosive mechanisms. But, do the higher, that is, flower-bearing plants also develop splash guns at all comparable to those of fungi, liverworts, and mosses?

Most Evening Primroses develop their seeds in a somewhat elongate brittle fruit that splits open lengthwise when it is dry and allows its seeds to fall where they will, or possibly to be scattered slightly by the sudden splitting open of the fruit. But there is one kind whose fruits remain closed when they are dry; only when such fruits are wetted by rain do they open out, and when they do, the lobes of the fruit fold back to form a cup. From that cup, seeds are spattered out by raindrops in much the same manner as spore-cases are splashed from the cup of the Bird's Nest Fungus.

There are many other examples of splash guns and no doubt many more will soon be recognized. It is really not surprising that this should be. To us, the use of the force of raindrops to accomplish one of the most important processes in the life of any living being, namely reproduction and spread or dissemination, seems to be a good idea. In the course of evolution, splash dissemination seems to have been developed in almost all groups of plants except the algae; but, after all, algae grow *in* water and we would hardly expect them to need splash guns.

Soil Drama

Out of the first dawning of agriculture must have come primeval man's interest in the soil beneath his feet. The earliest creations in graphic art reveal his overwhelming concern with those wild animals which were his contemporaries; they depict especially his activities as a hunter, for mainly by eating meat could he survive. But the hunt often must have been unsuccessful and meat often must have been scarce. Fruits, when they were to be found, as well as nuts and roots occasionally must have been a necessary part of man's diet, although even these constituted an uncertain supply, for fruits might be plentiful in any one season or year but absent in another. Then slowly, perhaps by chance, perhaps by some sluggish mental process, man came to understand that there were distinct advantages in husbandry. It was doubtless less exciting to keep animals near to him in domestication than to chase them over the hills and through the forests, but domesticated animals did provide a reasonably constant and certain supply of meat. Plants, too, provided a more readily available food supply when their seeds were scattered to grow close at hand than when they grew by chance far from where man had his primitive dwelling.

Beginning with the simple realization that there was good ground and poor ground, primitive man's resourcefulness must have been taxed in his efforts to improve poor soil so that it could yield crops bounteous enough to feed him. Out of such efforts there grew in time a vast knowledge, at first of a 'cut and try' sort, but ultimately to become an important science – an essential supporting girder in the basic framework of civilization. At first, man's knowledge of the soil was passed on to his inheritors as hearsay; soil management was by rule of thumb: today, soil science is a vast accumulation of knowledge whose frontiers are being extended by the most highly sophisticated techniques and sensitive complicated apparatus of modern science.

Our knowledge about the 'dust' from which we came is not only the concern of geologists who trace the evolution of soils from the seemingly unbreakable rock to its ultimate dissolution, or of chemists who identify the components of soils and study the influence of these components on soil fertility, or of practical soil scientists who enable man to obtain the utmost crop yield from his gardens and tilled fields. Soil is also very much the concern of biologists, for it is the domain of a very large proportion of the world's living things. Here dwell not only the green plants that are rooted in it or the easily visible animals that abide in its dark recesses (earthworms, insects, and many more) but also an entire microcosm of organisms so small that they are revealed to our sight only by microscope. We can appreciate that there may be many reasons for the biologists' interest in soil if we realize, on the one hand, that the soil hides many micro-organisms that cause diseases of crop plants, or on the other hand, that it hides certain bacteria that act upon nitrogen-containing substances in such a way as to make available to our crop plants much of that nitrogen which would not be available had it not been processed by the bacteria. Part of biologists' interest in soil micro-organisms might be considered to be esoteric: what are the organisms, what are their life habits, what peculiarities of structure or of biochemical activity do they display? But who knows? Even the most esoteric interests may eventually yield much information of importance to the welfare of mankind.

Writers of popular books (and they are legion) about micro-organisms present in soil usually report counts and estimates of the numbers of bacteria, protozoa, fungi, algae, nematodes (which are microscopic worms), and many other kinds of minute living beings that exist in an ounce or less of soil. Such figures are impressive in a way because they commonly run in the millions; but who can picture a million, or several million bacteria or fungi, especially since we cannot see most of those that may be present in our small sample of soil, even if we employ a moderately powerful microscope? Perhaps, therefore, we need to pause for a moment to consider how biologists go about finding and counting the ultra-Lilliputian denizens of the soil world.

Consider first the bacteria. If you were to place a few granules of soil on a small piece of glass, and then to add a drop of water – which makes examination under the microscope easier – and study the preparation even under a magnification of a thousand times, you would probably be able to see only a very few bacteria, or you might not see any. If the soil had been taken from the bottom of a stagnant pond, you might – just might – catch a glimpse of one kind of bacteria, and that because they are quite large as bacteria go; these have a distinctive corkscrew shape and they move with a jerky, twisting motion when free to move in water. Again, if the soil had chanced to have been in contact with some old wet hay, you might just see another large (!) rod-shaped bacterium because of its size and mobility. But probably you would not see most of these bacteria either because they are too small to be seen as individuals using ordinary microscopy or because they were present in a rounded-up resting state which you might not recognize as bacteria. And you would say 'A few perhaps, but millions in a few grains of soil ? Never!' You might justifiably wonder about the basis for biologists making such an assertion and perhaps conclude that they are given to what grammar books used to call 'hyperbole for the sake of emphasis.'

There are, however, several means of getting around this

difficulty of counting. One of these makes use of the centrifuge, a container of sorts which spins rapidly on a motor-driven axle. Particles in water in the spinning container are subjected to centrifugal force and tend to be forced to the side of the container. From there they can be collected; particles of various size and density classes can be separated, and it is thus possible to 'skim' out bacteria much as one can separate the cream from milk in a dairy separator. To count what has been collected remains a problem, however. The simplest means of accomplishing that is by diluting the suspension of bacteria. Starting with a known volume of the dense suspension, it is diluted in successive or serial operations 1 to 10, 1 to 100, 1 to 500, and so forth. A drop of each serial dilution is then examined under the microscope, and in one particular suspension there are so few bacteria within the optical field (that is, the area in view under the microscope) that they can be counted. Since the rate of dilution and the volume of liquid are known, one can calculate approximately how many bacteria were present in the entire original mass of soil.

The method is valid, of course, only provided the bacteria are large enough to be seen without extreme magnification, although, by using a dilution technique, with or without first centrifuging, it is also possible to account for the exceedingly small bacteria that could not be seen directly under the microscope. What is done is to 'culture' the bacteria; that is, to provide them with the proper nutrient materials so that they can grow and multiply. If we omit consideration of details immaterial in our context, the operation may be described thus. Small flat glass dishes are provided, each containing a solidified jelly (known as agar) to which basic nutriments have been added as food for the bacteria. The bacterial water suspension, which may also contain other micro-organisms, is placed in the nutrient-containing dishes, the particular dilution used being that which is estimated to disperse the bacteria widely and not provide too many in each plate. Each bacterial cell grows eventually into a mass of bacterial cells, each mass called a 'colony.' A colony of bacteria is a large enough object (if enough time is allowed for growth) to be seen with the unaided eye. So, by counting the colonies which develop on each plate, we can obtain a count of the number of individual bacterial cells that produced the colonies. If you could see a large dense crowd of people from some distance (as from an airplane), you probably could not count them accurately. If they dispersed widely and each person could establish a village (a colony) of people, you probably could count the villages from an airplane.

One source of error in such a method of counting is obvious: two or more bacteria might stick together in the diluted suspension and together give rise to only one colony instead of two or three colonies. Such an error, however, would give a number for the total of bacteria present in the soil which would be smaller than it should have been; the error would have led to a conservative count.

This technique of dispersing the micro-organisms of the soil has many uses other than the mere counting of them. A single spore of a fungus in the soil is almost impossible to identify, but when it grows on our dilution plates into a mass of mycelium, and especially if the latter produces fruiting structures which have a characteristic form, then we can identify the fungus.

With slight modifications depending upon the kinds of micro-organism being sought and their particular food requirements, the 'plating out' method can be used for obtaining and studying a large proportion of the bacteria, fungi, algae, and other living inhabitants of the soil.

This has been a long harangue, perhaps tedious. Anyone who tries to evaluate a statement given as fact, especially in an unfamiliar area, faces the difficulty of the necessity of understanding (to some degree at least) the evidence – in this case the techniques and reasoning involved.

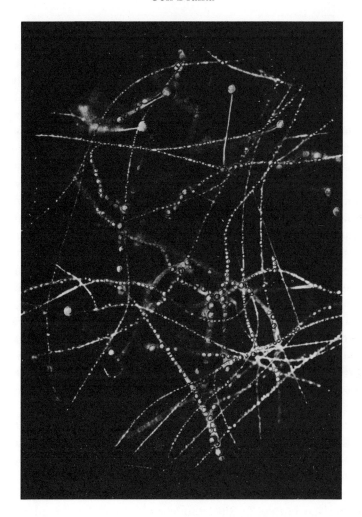

12 *Mucor:* This glittering fungus jewelry is actually a mass of threads or mycelium of a common soil mould bedecked with drops of moisture. Magnified about 200 times

Returning to the fungi themselves we shall examine a few that might have been growing on nutrient dilution plates, the cultures derived from the soil. One kind of fungus which is almost certain to be present, even abundant, is seen in Figure 12; it bears the name *Mucor*. There are many kinds (species) of *Mucor* and it is but one of a large group of moulds of which the common bread mould *Rhizopus* is the one with which the average person is most likely to be familiar. Moulds such as the two named develop as rather conspicuous cottony masses on old damp bread and decaying vegetable matter of all sorts. The cottony mass is, of course, the mycelium or vegetative part of the plant. When *Mucor* mycelium is growing in closed glass plates as is the one in Figure 12, the individual threads of the mycelium are frequently adorned with water droplets and present the appearance of chains of pearls or remind us of a spider's web bedecked with dew. The jewels on the fungus threads, however, are usually not pure water but contain various substances exuded by the fungus. Some of these mould products are of importance to mankind. For example, the common bread mould can be used commercially for the production of fumaric acid and is involved in some stages of the manufacture of the drug cortisone. Other moulds produce significant amounts of alcohol, others of commercially important acids such a citric acid, oxalic acid, and other chemicals. Many of these moulds cause much destruction of fruits and vegetables, especially in transit, and a few kinds, beautiful and harmless though they appear, cause diseases of humans. Some in the latter category affect the internal nervous system, and may occasionally have fatal consequences.

Mucor and its close relatives reproduce, most frequently, in a non-sexual process by the production of prodigious numbers of spores which are borne in beautiful spherical shiny spore-cases. Each spore-case surmounts a tall stalk and the whole spore-bearing apparatus reminds one of an old-fashioned black-headed pin because the spores themselves are often black in colour and impart a very dark appearance to the globe in which they are developed. A few such pin-like *sporangia*, as they are called, can be seen in the upper third of Figure 12. When the spore-cases are mature, their walls become brittle and, as they crack open easily, the spores are released to be dispersed far and wide by the wind. As for their fate, some may land eventually on soil where, if it is moist, they develop into more mycelium within the soil and play their unseen role on that dark stage: some, too, may land on the surface of your best grape jelly.

The names which have been given by biologists to living organisms are Latin names for the reason that it has long been a language of universal use, a sort of standard means of communication when much else would fail. Much has been written, and perhaps some of it has been read, about why these official names are given in Latin rather than English or Czech or Bengali and no more needs to be said. Although most Latin names are really not difficult to pronounce, the rules for pronunciation being more regular and logical than those of many other languages such as English, it must be admitted nevertheless that such a name as *Schizosaccharomyces*, which is a perfectly harmless yeast plant, is scarcely euphonious. But very few fungi have been blessed with a name which has as mellifluous a sound as that of the lovely mould *Circinella* which is shown in Figure 13. The name is a modification of a Latin adjective *circinalis*, which might come out in English as 'coiled inwards from the tip.'

And what beautiful coils or curves are those of *Circinella*! The fungus bears spore-cases or sporangia at the tips of long stalks much as do *Mucor* and *Rhizopus* and is mostly found in soil or growing upon decaying vegetation. Except for giving names to the various species of *Circinella*, biologists seem not to have discovered anything of an unusual or 'important' nature about it and so it is

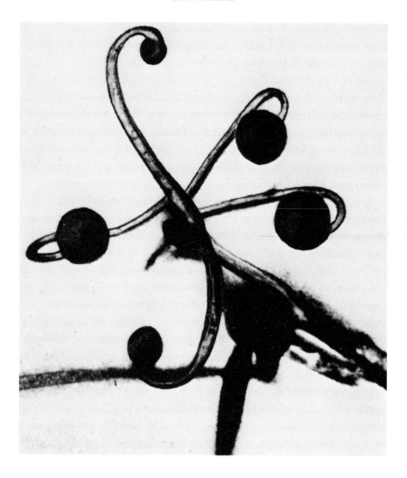

13 *Circinella:* The imaginative abstractions of an artist do not approach the natural beauty displayed by the spore-producing stalks of this soil-inhabiting mould. Magnified about 800 times

presented here, unheralded, as a lowly soil-inhabiting organism, little seen but surely one of the transcendent beauties that reward those who care to look about them.

The familiar expression 'slippery as an eel' must surely convey to most people the notion that to catch one with bare hands is not a simple matter. Whether or not a similar degree of 'slippery-ness' is possessed by the microscopic soil-dwelling animals called *eel-worms* is, of course, unknown; however, if one can draw any conclusions from watching them under a microscope, they may be very slippery indeed. These very tiny elongate earth creatures taper to a point at each end; they move in a writhing and twisting manner much as a snake does, and with astonishing rapidity. One feels instinctively that they would be difficult to apprehend were they within our size range rather than being only about 1/32 to 1/16 of an inch in length and usually not over a third of that in diameter at the middle. However difficult and improbable it may appear, it is nevertheless a fact that there exist, in the very soil we have been examining, large numbers and a considerable variety of fungi that gain their livelihood by virtue of their truly remarkable adeptness at snaring or trapping eel-worms.

Long ago, a few microscopists had described seeing eel-worms apparently caught, writhing and expiring in various loop traps and snares of certain kinds of filamentous soil fungi. Naturally enough, many biologists were not convinced and considered that, even if the eel-worms did sometimes get caught in fungus lassos, it was an accidental occurrence. But considerably later it became clear that the soil drama involving hapless eel-worms was a matter worthy of serious study. It was found possible to obtain pure cultures of the trap fungi and to demonstrate that at least for some of them no loop traps were formed by the fungus unless some eel-worms were added to the pure culture of the fungus. When plenty of trap fungi and eel-worms were available by culturing them, it

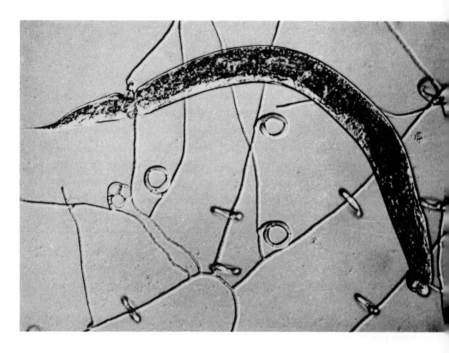

14 *Arthrobotrys anchonia:* The three-celled snares of a soil fungus close and
trap an eel-worm in a regularly occurring drama among microscopic
organisms in the soil. Magnified about 200 times

was possible to observe the fascinating relationship repeatedly;one could study the structure of fungus loops and watch the behaviour of trapped eel-worms (Figure 14).

Several quite different groups of soil-inhabiting fungi have developed some means of attaching themselves to the writhing slithery eel-worms found in soil. In some of these fungi, certain of the filaments that compose them take the form of an extensive system of loops or nets in which a cavorting eel-worm may easily become enmeshed. Getting free from the network, however, is an entirely different matter, for the nets are made highly adhesive by the excretion by the fungus of a very special kind of glue-like material, one that is quite remarkable for it is not dissolved by water.

When an eel-worm has, inadvertently, poked its head (or tail) into the sticky loops, it is improbable that there will be any reprieve for the reckless worm. The fungus network holds fast and the worm must yield its store of carbon, nitrogen, and the other staples necessary for all life; the fungus threads proceed to penetrate the body of the trapped worm and digest the contents of the animal's body.

Some eel-worm catchers bear sticky knobs that easily become attached to the victim, some others bear snares that are simple rings composed of several cells of fungus filaments. The most intriguing trap of all, however, consists of a ring that can contract very quickly when an eel-worm has poked its head through it. Each ring is made up of three cells (Figure 14); in the 'unsprung' trap these have about the same diameter as those of other filaments of the fungus but, when a worm has wriggled into the trap, the three cells of the ring almost instantly swell inwards (towards the centre of the ring) and a tight constriction is thus formed around the worm's body, so tightly closed indeed that the worm almost never escapes. Unswollen rings and one swollen ring can be seen in the photograph as well as the two rings that have snared the worm, one at each end.

Well may we marvel at the agility of this wild-west microscopic mould and ask how the ring trap operates. Biologists who have studied constricting rings have reported several observations that have helped to explain its operation even if all the necessary facts are not yet known. First, it has been observed repeatedly that swelling of the three cells takes place very quickly, in about 1/10 second in fact. Second, one can spring the trap with a fine needle, which suggests that it is pressure that is essential, not the chemical nature of the worm's body surface; but pressure from a needle must be applied to the inner surface of the ring, not the outside. Third, there is a most obvious increase in the volume of each cell when the trap is sprung; rapid change in the permeability of the living protoplasmic membrane of the ring cells take place which allows a rapid intake of water, which in turn causes the ring cells to swell. Observers have reported that a writhing worm may (rarely) free itself by tearing the fungus network; but, when this occurs, the worm still bears the closed ring, which means doom because the ring cells ultimately grow into its body anyway: thus, the fungus 'gets its worm' and thus it gets its food; continued life is assured for the little soil fungus.

Biologists generally dislike being asked the question: 'Interesting yes; but what good is it?' They dislike it not only because it is often difficult to answer, but because the question is highly loaded in favour of man; it is anthropocentric. No doubt the question will arise nevertheless, so here is an answer of sorts. If by 'good' is meant beneficial to mankind, there may be something to be said for the activities of the fungi that prey upon eel-worms. Although most of these are what is called 'free living' and obtain their nourishment from dead organic matter in soil, quite a large number of kinds cause diseases of the roots of crop plants to such an extent as to be economically important. It seems highly probable that eel-worm-catching fungi may reduce the eel-worm population significantly, especially at certain times under particular conditions; if the fungi do reduce the population of parasitic eel-worms then no doubt the fungus traps might be considered a boon by mankind. There is no doubt either that eel-worms themselves, if they could be consulted, would have a different viewpoint!

And the three blind men? One of them feels our lump of soil and correctly identifies it as soil which he does not care to tolerate adhering to his person or find mixed with his food. A second crumbles the soil in his hand, smells it, and declares that it smells 'rich' and 'fertile' and should produce good crops. A third, when he has been obliged to listen to our story perforce because he has no light vision, pauses meditatively and says humbly: 'But verily an entire world apart, a world of drama and of surpassing beauty where it must surely seem there could be neither.'

The Sphere Thrower

Confronted with the small objects of Figure 15 – just *how* is left to the reader – one of the three blind men might have thought of a pearl, another of a tiny star-shaped flower; and if one of the three had been born in recent years he might have thought of bubble gum, an image mercifully absent from the minds of the two older sages!

Again each image has some value merely as description, but not one is likely to provide a clue as to why the little fungus should bear the name *Sphaerobolus* – the 'sphere thrower.' And what a champion sphere thrower it is! Neither the discus thrower of ancient Greece nor his modern counterpart could hurl the discus to a distance of more than a few times his own height, but *Sphaerobolus* hurls its tiny cannon-ball projectile to a distance several hundred times as great as its own width.

This tiny champion fungus, seldom more than one-sixteenth inch in diameter, appears usually (and fairly frequently) in considerable numbers as spherical whitish or yellowish-pink globes (Figure 15) growing on old manure, old dead wood, or even on old boards in damp places. It is almost of world-wide distribution in temperate climates and occurs also in cool, moist regions in the tropics. In its general form, as well as in many technical details of its structure, *Sphaerobolus* resembles the puff-balls and is in fact related to them. But whereas they liberate their spores either by the action of wind or through the battering of raindrops which de-

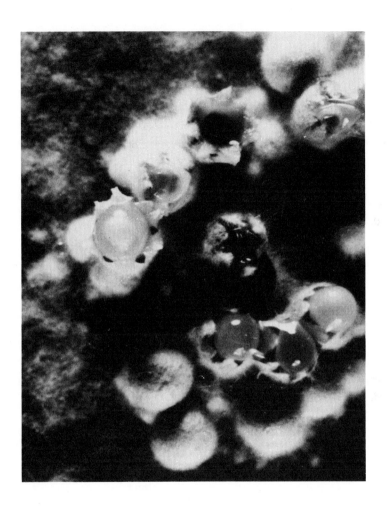

15 *Sphaerobolus:* This tiny puff-ball-like fungus throws out a spherical mass of spores by suddenly turning one of its inner tissue layers inside out. The dark spore mass can be seen in the open fungus mouth near the top; lower in the photograph are the pearly spheres of the everted inner layer. Magnified about 10 times

press or indent the papery wall of the spore-case and 'puff' the spores out from the open end (in an atomizer kind of action), *Sphaerobolus* shoots out its spores by means of an amazing and highly effective kind of catapult mechanism. Moreover, whereas puff-ball spores are distributed in the air as separate particles like motes of dust, the spores of *Sphaerobolus* are shot out of the fungus together as a single solid mass – a ball containing a very large number of spores enclosed within a firm case which is highly adhesive externally and because of which feature the spore mass becomes glued to whatever object it may strike (in nature, the latter is usually a leaf or a blade of grass). The adhesive projectile, containing spores, remains attached to the target until the blade of grass (or whatever) is eaten by a herbivorous animal, thus ensuring further dissemination or spread of spores of the parent fungus. Spores germinate eventually in the animal's feces wherever these may be dropped and, sometime later, we find *Sphaerobolus* fruiting anew on some old manure or on some old boards nearby.

The violent discharge of the cannon-ball-like mass of spores is accompanied by a distinctly audible 'ping' sound, and so forceful is the ejection that a sound may also be heard when the projectile strikes a solid body. More than one photographer has been struck in the face by the spore masses while bent close to the little fungi in an endeavour to focus his camera!

It is merely a matter of patience to observe the little fungus 'mortars' in action, provided one has found a good supply and keeps them moist; if they are allowed to dry, they cease firing. One can measure the little projectile (about 1/20 inch in diameter), weigh it, observe that it is sticky or adhesive, measure the distance to which it can be hurled (a 6 to 10 foot throw has often been measured), and so on. However, to understand the mechanism by which the Lilliputian mortars operate, we need to examine the fungus more closely. Using only the unaided eye, we observe that the entire fungus is a fluffy, pinkish, closed ball when very young (see Figure 15). Each closed ball is securely fastened to whatever it may be growing upon and cannot easily be detached without breaking the little fungus gun. It is clearly well anchored to the substrate by part of the mycelial threads of which it was formed. Later, an outer layer splits off radially and opens out stellately, the fungus then appearing somewhat like a partially peeled orange. At this stage of development, moreover, we can see the exposed contents, a yellowish or pink sphere (the spore-mass) which is moist and shiny and which gradually deepens in colour to brown or almost black. If we have the patience to wait for events, we can observe the violent discharge of the spore-mass and should soon realize that we are on the trail to understanding what is happening when we have become aware that the appearance of a discharged gun is very different from that of an undischarged gun. After the discharge, the pink or brown spore-mass is gone and the centre of the little mortar is occupied by what seems to be a pink shiny bubble, for all the world like a wad of blown-out bubble gum ('like a pearl' would be a more polite description, but a less apt comparison). We might, moreover, come to the conclusion that something from the inside of the mortar (in addition to the spore projectile) has popped out; something seems to have turned inside out.

Farther than this in solving our problem we cannot go unless we are able to examine the structure of the fungus by means of a microscope. We must now contrive to obtain very thin slices of *Sphaerobolus* not only when it is very young but at successive later stages of its development right up to that at which it is ready to fire off its projectile. (Exactly how this preparation of thin slices is accomplished is not important to our understanding of the structure of the fungus, but it is very important for the biologist who is trying to elucidate the microscopic details.)

By studying the thin slices under a microscope, one can discover that even in a very immature specimen the mycelial filaments of which it is composed are already organized into six distinct layers or regions, each different in structure and organization from the others and each eventually to have some role in the operation of the fungus mortar gun. Much careful study was required to discover how each layer continues to develop into the parts of the 'machinery' which finally comprise the mature gun. Omitting the details of development, we can still easily comprehend that, at maturity, the body of the fungus comprises mainly two layers of tissue. At one point in development, the spore projectile is completely surrounded by these two layers but, at maturity, when the outermost covering layer of the entire fungus splits and peels open, the two principal layers have become almost completely separated from one another; they do remain attached, however, at the tips of the points of the outer layer. We can visualize the structure at this stage if we think of two cups nested one within the other but attached in places around their lips. Lying within the now open inner cup is the spore-case projectile, surrounded by a sort of lubricating liquid which has been produced as the result of the disintegration of filaments of one of the six layers mentioned.

The innermost layer, on which the projectile rests, undergoes chemical changes. It is composed of long, more or less cylindrical cells in which pressure is steadily increased by the chemical changes. Finally pressure in this 'eversion layer' reaches a point at which the stress that has been developed must be relieved somehow; so, the eversion layer 'everts' or turns inside out and the loose projectile is catapulted out of the open cup. The eversion layer, remember, is still attached at a few points to the outer part of the cup, and so it does not fly away with the projectile but remains behind on the outside of the cup as a spherical pearly object, as we can see in Figure 15. And thus the spore-mass cannonball is hurled away from the tiny mortar. Because the spore ball is surrounded by an adhesive liquid, it becomes securely glued to whatever object it may strike. There it remains and if its resting place happens to be grass or other herbage, so much the better; for then – like the projectile of another fungus gun *Pilobolus* already examined – it can be transported to still more remote sites when the grass is eaten by a wild or a domestic herbivorous animal.

To some extent, though not with the accuracy of the *Pilobolus* gun discussed earlier, the mortar of *Sphaerobolus* is able to shoot its projectile towards the light. During the growth of the spherical fruiting structure, its development is influenced by the direction from which the strongest light is received – though exactly how the influence is exerted is not completely understood – and the little fungus mortar becomes inclined towards the light as it matures. When, finally, the spore-mass projectile is shot away, it therefore is able to make the longest possible journey and to adhere to herbage out in the open, as free and clear of other obstacles as possible.

In this respect, the aiming of the *Sphaerobolus* gun achieves the same objective as does that of the light-directed gun of *Pilobolus*, although these two fungi are far apart in the matter of their position in the scheme of classification of the fungi in general acceptance. *Pilobolus* is closely related, in basic respects, to the common Bread Mould (*Rhizopus*); *Sphaerobolus* is closely related basically to the puff-balls – a group considered to stand, in systematic arrangement and considered in terms of evolution and complexity, far above the moulds. The two very different kinds of fungi have nevertheless solved the problem of spore dispersal in what is basically the same manner: both produce spores in a single mass which is adhesive, is propelled towards its target, and is aimed by making use of light energy.

We have already considered a comparable example in the water-guns, where the force of rain splash is made use of by such unrelated plants as fungi, liverworts, mosses, and flower-bearing plants. An otter and a trout swimming side by side in a stream both accomplish the major part of their forward-propelling swimming by a sinuous or bending motion of the whole body, though an otter is a much more 'advanced' type of animal than a fish. The lesson in these and countless other examples is that basic problems which are of prime importance in the struggle for survival often have been solved in basically the same manner, though the organisms we are comparing may be quite unlike structurally and taxonomically. In the long course of evolution, the inherent and accumulated potentialities of each kind of organism have enabled it to overcome whatever obstacles are presented by the problems of survival.

The Insect 'Itch'

There seem to be very few places on earth where some kind of fungus cannot find enough nourishment for its needs. Green plants, fruits and vegetables of all kinds, old wood and soil – all these appear as rather-to-be-expected sites for fungus growth. A few less-to-be-expected substrates on which fungi thrive because they are able to obtain some nourishment, though it may be meagre, are old feathers, leather, bones, paper, and cloth. A few moulds can survive and even grow fairly well in dilute solutions of certain chemical salts. These moulds in the old days used to crop up on the chemist's shelves (we would say druggist's today) in seldom used bottles of solutions for making up prescriptions, and they were called 'bottle imps.'

One of the most unusual of all places where fungi grow harbours some of the most unusual of all fungi, the Laboulbeniales. (We must accept that name as best we can, for there is no other.) This is a large group of many hundreds of species of microscopic fungi which grow almost exclusively upon the bodies of insects, a very few kinds only having been found growing upon mites. Note the word *upon*! The strange tiny fungi do not grow *into* the insect body to any considerable extent as do some other kinds that are parasitic upon and destroy many kinds of insects. The Laboulbeniales seem content to remain attached to the outermost parts of insects' bodies. Ordinarily, the hardest part of an insect's body is the outermost part which entomologists call the exoskeleton, a situation contrasting with that in large animals where the bony framework lies on the inside of the body. One would think that there could not be much to eat, even for a fungus, on the hard exoskeleton of a beetle; the problem, therefore, of how these curious fungi nourish themselves is also one that still puzzles biologists greatly. The problem is not lessened by the knowledge that the tiny fungi seem to have little or nothing comparable to roots that might penetrate deeply or extensively into the insect body; moreover, the Laboulbeniales have not, up to the present, been grown in artificial culture, that is, in some kind of a nutrient decoction such as any one of the many that will support the growth of a wide variety of other fungi. At present, it is not fully known how the Laboulbeniales obtain their food. Those who have studied these fungi intensively are inclined· to believe they are parasites and that the parasitism is an obligate relationship, but there is only circumstantial evidence to support this view.

From Figure 16 we can obtain some understanding of the appearance and structure of one species of the Laboulbeniales. In size, the entire fungus body is rarely over 1/32 of an inch in length, and usually much less; it is (for most species) too small to be seen in detail with the unaided eye and is barely discernible under a stereoscopic microscope magnifying about twenty times. The first point of unusual interest about our tiny fungus is that it lacks the extensive network of threads, the mycelium, that characterizes most fungi. In fact, the entire fungus is made up of a very small number of cells, usually a couple of dozen at most and, for some kinds, fewer still. In most species these cells, which are mostly brownish or even shiny black when seen under the microscope, comprise four main parts or regions of the fungus body: there is a stalk of sorts, the lowermost cell of which is called the 'foot cell,' which fixes the fungus to the insect's body; attached to the stalk or receptacle are various appendages, some of which produce the male cells for reproduction and some of which – often

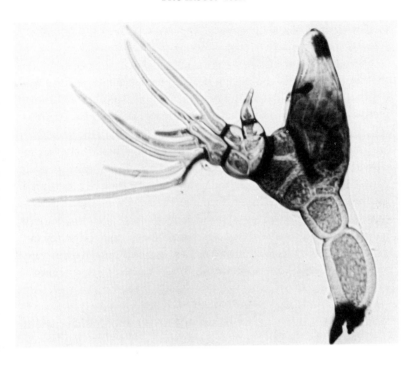

16 *Laboulbenia:* A minuscule fungus consisting of only a few cells. Black
basal cells are the foot or attachment; six stalk cells bear the male reproductive branches at the left; the female apparatus is at the right. The
fungus grows upon the external parts of insects.
Magnified about 600 times

pear-shaped or egg-shaped – are the female parts of the reproductive apparatus. And that, essentially, is all there is.

The relative simplicity of the fungus body, at least in terms of the number of cells composing it, makes it all the more remarkable that the Laboulbeniales have been able to develop an almost incredible variety of the most bizarre forms. Well over 1500 species are now recognized, largely the result of the patient searching and meticulous investigations of the late Dr Roland Thaxter of Harvard University, whose published monographs (nearly 2000 pages) constitute one of the most outstanding classics concerning our knowledge of the fungi. Professor Thaxter's detailed descriptions are accompanied by many dozens of large plates of superb hand-drawn illustrations that have never been surpassed as accurate and artistic records; they represent a signal and inspiring example of selfless lifetime devotion to the cause of science. Naturally there are many who will consider that this man's time and energy would have been better devoted to something more 'useful'; but *that* has been said on one occasion or another about most of the creative and enduring efforts of mankind.

In seemingly endless variety the Laboulbeniales appear before us in Thaxter's pages. Inadequate though words are to portray what we see, imagine a weird science fiction garden of plants from another planet, but reduce them all to microscopic or almost microscopic dimensions. Here is a simple one that looks like a Saguarro cactus with ungainly arms. Another resembles a couple of barrels one on top of the other and having long feathers attached. The appendages of still another remind us of the antennae or feelers of a large male moth, or perhaps an ostrich plume. In some kinds, the receptacle suggests a broad snake with patterned segments, another a broad vase filled with bananas, all on end.

Such descriptions are ridiculous, but they may convey to some people a better visual conception of these bizarre fungi than the actual technical descriptions as given in Thaxter's monograph, of which the following is but a short example:

> Perithecia large, sub-cylindrical, terminating rather abruptly and symmetrically rounded below the short, clearly distinguished, broadly truncate apex, the lip cells of which end in papillate enlargements, while one of them produces posteriorly a clavate outgrowth bent abruptly upward near its base ...

The striking resemblance of many of the Laboulbeniales to parts of insects' bodies – antennae or 'feelers,' parts of legs, parts of mandibles or 'jaws,' hairy eyes, and so forth – has an interesting relationship to the discovery and recording of many of these fungi, not so long ago, by entomologists who wondered if the fungi might be extra parts or abnormal parts especially of such hard-bodied insects as beetles.

Whatever else can (or should) be written about these strange fungi, they afford one of the most astonishing examples of structural variation manifest among multicellular organisms, consisting as they do of so few cells that one might think of them as having very little to work with. It is true that even bacteria, which are one-celled organisms, and some one-celled algae do display great variation but, for the most part, variation in the latter has not resulted in the superabundance of unbelievably beautiful body forms – in many instances grotesque. One cannot escape the temptation to wonder whether or not any of the strange contraptions developed by the Laboulbeniales have any survival value or are of any beneficial significance for the fungi themselves. When one considers that Professor Thaxter devoted forty years to the patient and painstaking study of the Laboulbeniales, it may be surprising to learn that some of the most important aspects of the lives of these fungi are still unknown (surprising to many persons, that is, but not to biologists, who are quite accustomed to the way in which nature often persistently hides some of her secrets). Each of our little fungi begins its development as a minute spore, called an *ascospore* (because it is developed in a sac-like structure called an *ascus*) commonly composed of two cells. The spore is surrounded by a sheath of very adhesive material which enables it to stick firmly to an insect's body. In growth, almost at once, one of the two cells becomes the fungus foot and penetrates sufficiently into the insect exoskeleton so that the fungus remains anchored. The other cell produces more cells and gives rise to the remainder of the fungus body, namely the stalk and its appendages, chiefly the female element and the male, but also some of the other curious sterile, ornamental (?) parts. The male elements are small non-motile cells either produced by the segmentation of long branch appendages or extruded from small flask-shaped bodies. It is not known exactly how or when male cells become attached to a special receptive stalk called the *trichogyne* (they are often seen attached to it) which protrudes from the female structure when the latter is mature. Nuclei from the male cells migrate via the trichogyne into the long flask-shaped female body and fertilization is effected, stimulating the production of ascospores within the female part. It has been observed that, for some kinds at least, ascospores are ejected from the female body if pressure is applied to it, as by a needle. Supposedly, therefore, sticky ascospores must be 'squirted' from one insect on to another when the insects are in direct contact as, for instance, during copulation. In some species, the fungus bears a long, curved, tusk-like structure solidly attached to one side of the female part and this may act, when touched, as a trigger (Figure 17) which transfers pressure to the female structure, thus causing ascospores to be ejected.

The transfer of sticky ascospores from one insect to another (only, or mostly) during mating of the insects may explain the often-repeated but almost incredible observation that many of the

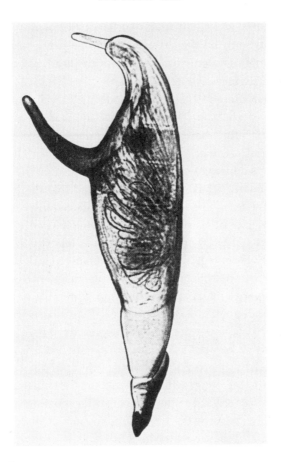

17 *Dioicomyces spiniger:* Another member of the Laboulbeniales suggests a strange tropical bird. This minute fungus grows upon the exoskeleton of beetles. The 'beak' is the trichogyne or female receptive organ. The curved appendage at the left probably functions as a trigger to discharge the spores that are formed in the series of tubular bodies within the body. Magnified about 450 times

The Insect 'Itch'

Laboulbeniales grow only upon certain parts of any particular insect's body: only on the left wing cover of a beetle, for example, or only on the right; only on the back of the head, or between the eyes, or on front legs but not back legs.

As for the effect of these fungi upon the insects that harbour them, little can be said. Thaxter himself and many reputable students since his time have reported that infected insects appear to be 'restless' in comparison with non-infected ones and that infected insects appear to make some effort to scratch the tiny fungi from their bodies; Thaxter himself suggested that the fungi cause a sort of mild dermatitis (skin disease) of infected insects.

Biologists, like other scientists, are schooled at an early age in the tradition of objectivity; they must *try* to report what they observe *as* they observe it, refraining from taking a position of interpretation which reflects only their own attitudes as humans. Such an ideal is very difficult to achieve; and admirable and virtuous though it undoubtedly is, there may be no real harm done (occasionally) by a slight deviation from the prescribed path – provided it is honestly recognized as a deviation. Therefore, it is a mild but perhaps diverting heresy to wonder if insects *itch* when they are infected with the fungi which we call the Laboulbeniales. This idea has been expressed in the amusing little poem usually quoted anonymously, but apparently written by an English physicist-mathematician, Augustus De Morgan:

> Great fleas have little fleas upon their back to bite 'em
> And little fleas have lesser fleas, and so *ad infinitum*.
> The great fleas themselves in turn have greater fleas to go on
> While these again have greater still and greater still and so on.

Mixing our metaphors, it is obvious that our blind men are in deep waters. Even if the fungus bodies of the Laboulbeniales could be enlarged so that each member of our panel of judges could feel the plant body with his own hands, the explanatory descriptions that would ensue might well be amusing, but no one of these would alter the bewilderment we feel in the presence of such a surfeit of forms so strange to us. And, at this stage of our knowledge of these unusual fungi, little if any enlightenment is likely to come from a question involving the word 'meaning' (significance, purpose, use, or what you will) if we reflect upon the question sufficiently to ask: *meaning?* – for the fungus, for the insect, or for ourselves?

The Stone
That Grows

Some of our most poignant nostalgic emotions arise from contact with such simple elements as the crackle and glow of the hearth fire and the odour of freshly ploughed soil. For too many dwellers in twentieth-century concrete catacombs, the nostalgia may not be derived from vivid recollections of special circumstances in their own past; for too many of them, fire is an insurable risk and soil is that which necessitates a visit to the nearest dry-cleaning establishment. But whether it is an emotion of realistic and recent origin or whether for all mankind it is an atavistic nostalgia that harks back to ancient hearths and the first ploughed lands, the emotion is one of our blessings.

Biologists doubtless also love the odour of fresh soil, but they are inclined to consider other aspects of it as well. If questioned on the matter of its aroma, a biologist would point out that only certain kinds of microscopic soil-dwelling micro-organisms produce chemical substances which are responsible for that special odour. He would then hasten to endeavour to share with us his own fascination in contemplating the soil as a living microcosm. Here, he says, dwell millions upon millions of organisms too small to be seen, or at least to be examined in detail, except through a microscope. And in this microscopic subterranean world, even as in the realm of larger living beings, the basic necessities of all life are striven for by all and acquired by some. The increasing competition for food and the inexorable drive towards reproduction lead to complex relationships among the soil denizens, relationships that are basically much the same as those that 'complicate' life for larger organisms.

Making up a large proportion of the living entities in soil are the intricate networks of the microscopic threads of fungi, collectively called mycelium. That which belongs to some of the simpler and more primitive kinds of fungi may pass through the entire life cycle unseen by our unaided eyes. From one generation of spores which will develop into mycelium which in turn will eventually produce new spores, such fungi remain, at all times, too minute to be detected except by microscope. In contrast, mycelia of other kinds of fungi generally believed to be more advanced or highly developed sorts, though they likewise spend most of their lives underground and unseen by us, at the time of reproduction build large and conspicuous reproductive structures such as the mushroom, the Coral Fungus, and many thousands of other forms. It would be very difficult, almost impossible in fact, for a biologist to show you the delicate mycelial threads of a mushroom growing in the soil of your lawn until and unless the time for reproduction had come; at that time the mycelium forms spherical masses of interwoven threads at ground level and large enough to be easily seen, especially when the spheres have enlarged rapidly and become full-grown mushrooms.

It is not surprising that many kinds of subterranean fungi, even large ones, are almost never seen unless they are accidentally revealed when soil is dug or ploughed. The truffles, for example, which are much prized as edible delicacies, are not revealed to covetous eyes until pigs trained for the task have sniffed them out through the sweet and characteristic odour truffles produce. In this remunerative pastime (remunerative for the pig's owner), the pig is then tethered while its owner unearths the treasure. If the pig is fortunate, it is fed a truffle now and then to maintain interest.

One of the strangest of all subterranean fungi, shown in Figure

18 *Polyporus tuberaster:* Resembling a large black stone when it is turned up in ploughed fields, the sclerotium of this fungus is a 'resting stage' or a storehouse of food on which the fungus can draw after a drought or when winter is over. It has often been called 'fossil pemmican.' About two-fifths natural size

18, is called the Canadian Tuckahoe. Unlike the truffle, it is found not in the beech and oak forests of France and Spain but in the parkland belt of Canada and the northern United States. The tuckahoe is most often found by farmers while ploughing, and occasionally by someone digging a garden. The farmer often sends his curious find to the nearest school teacher or to some other person who might be expected to know about curios. Whoever it may be that issues the pronouncement, the verdict is frequently as far from the truth as would be the verdict of any one of the three blind men. Out of charity we shall not ask for their opinions, largely for the reason that it is extremely doubtful that the Canadian Tuckahoe could even have been found in whatever country the three blind men of the old legend inhabited.

We can, however, with possible profit and interest take note of some of the guesses that are often made as to the identity (or perhaps the remote affiliation) of the tuckahoe. When it is fresh and moist it is soft and rubbery or even leathery. When it has dried somewhat, it is very tough in texture: to cut it through one is obliged to use a saw; and, when it has been sawed and one can examine its internal appearance, it is seen to be mottled (Figure 18, right). In the latter guise, tuckahoe has been identified as 'fossil pemmican.' When completely dry, tuckahoe may resemble a rock which is very hard and which may be as much as 10 by 16 inches in size and up to 10 pounds in weight. Many an amateur geologist or rock hound has been hard pressed in endeavouring to identify an old, dry tuckahoe as some kind of rock.

The Canadian Tuckahoe (the scientific name is *Polyporus tuberaster*) is a large dark mass, irregularly spherical or oval in shape and often slightly flattened. On the outside the fungus body is a dark grey colour, sometimes almost coal black though dull, not shiny. The external surface may be marked with lines or ridges resulting from its contact while still in the soil with a root, which may even be partly enclosed by the tuckahoe (Figure 18, left), and the external surface very commonly bears numerous criss-cross wrinkles. If we try to cut through a tuckahoe with a knife it immediately becomes evident that the internal part of the mass has grains of sand and even small stones embedded in it. Below the black crust, the interior part of the tuckahoe is a very dark olive green in fresh specimens. In older, dry specimens the interior is dark grey but appears generally lighter because dark-coloured areas are interspersed with dirty white areas; this gives to the interior a mottled or marble-like aspect (Figure 18, right).

When an old, dry specimen is soaked in water it can absorb as much as fifty per cent of its own dry weight but, even when softened in this way, it is too tough to be chewed. If boiled it becomes slightly softer but still not an object of likely palatability, which makes it difficult to understand why hearsay has it that a tuckahoe is edible; it might possibly be edible when very young and fresh, but certainly not otherwise. Whether or not the North American Indians ever used the tuckahoe as food is, therefore, an open question despite one of the names that has been applied to it, namely 'Indian bread.' However, it is recorded that the Indians did use the fungus as a poultice and for rheumatism, and that they called the tuckahoe 'ground medicine.'

And what about the epithet 'fossil pemmican' that has been applied to tuckahoe? Pemmican (the name is derived from the Cree word *pimikkan*), a staple of the diet of North American Indians, was made from lean buffalo meat or from venison which was cut into strips and dried in the sun. The dried meat was then pounded into fine bits, mixed with melted fat, and packed in bags made of buffalo hide. When pemmican was made by early explorers, dried fruit was sometimes added. Such compressed cakes contained much nourishment in small compass and were widely used on long voyages and even cached as an insurance for the return trip. How tuckahoe came to be known as 'fossil pemmican' can only be conjectured, but we can imagine that the thinking must have followed somewhat along these lines. On the exterior, a tuckahoe is black and its surface is marked by fine criss-cross lines or wrinkles, both features suggesting the remains of a bag or sack made of hide. The interior of a tuckahoe is of irregular texture, the mottled or marbled appearance suggesting meat. The tuckahoe is found underground, suggesting that it might have been hidden there or lost and buried accidentally. Finally, when a tuckahoe is old and dry it is as hard as stone, suggesting that it might have become petrified. From all of this, possibly, 'fossil pemmican'?

One of the most surprising characteristics of this strange 'stone' of the prairie parklands is that it will grow (Figure 19). Such a statement would not surprise many Canadian farmers who wrested a living from poor stony soil from which they were obliged to remove, each year, a fresh crop of stones brought to the

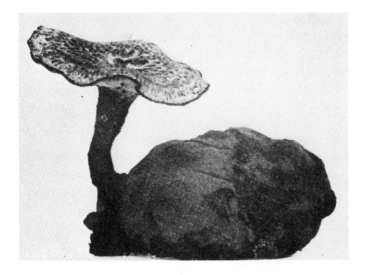

19 *Polyporus tuberaster:* The fungus stone of Figure 18 has renewed growth after an arresting drought or the winter cold and has produced its mushroom-like reproductive stage from materials stored within the stony sclerotium. About three-fifths natural size

surface by the heaving action of frost, and who often declared: 'these stones grow.'

Nor would the 'growing stone' surprise a biologist, for he is quite accustomed to seeing fungi perform such a trick. Many kinds of a wide variety of fungi produce very hard storage bodies or 'resting' stages which are known as *sclerotia* (singular *sclerotium*). A sclerotium is a hard, dense body, often roughly globose, which serves to keep the fungus alive throughout a dry season or throughout a winter when moisture and temperature conditions make active vegetative growth of the fungus threads (mycelium) impossible. A sclerotium itself is composed of mycelium, as are all parts of a fungus; mycelium of a sclerotium, however, is greatly modified from that which grows actively as delicate threads in the soil. Walls of the threads which make up a sclerotium are greatly thickened and they are very densely interwoven and compacted. Both these features give to the sclerotium its tough, horny, or even stony texture. Moreover the thickened mycelial walls and the compaction of the material, in addition to the presence of food materials such as oil and glycogen, make of a sclerotium a veritable storehouse from which the fungus can withdraw a great supply of food when the time for reproduction is at hand, and weather allows growth to occur. In most kinds of fungi, these storage bodies do not develop directly into vegetative mycelium when growth is renewed; rather, the sclerotia produce the structures necessary for sexual reproduction.* This is also true of the sclerotium of *Polyporus tuberaster.*

* Parenthetically, at this point, especially to aid the thinking of persons who may have had little contact with the basic concepts of biological science, it may be noted briefly that reproduction by means of a sexual process of some sort is of paramount importance for all living organisms. What it accomplishes, basically, is the random separation or segregation of the heritable characteristics of organisms in the sex cells they produce and the random recombination of these (when sex cells unite) to begin the next generation. The resultant offspring may therefore contain different chance combinations of heritable characteristics. The individual offspring thus may vary but not necessarily from one another in the heritable characteristics acquired from their parents. Since the total 'package' of heritable material inherited by any one offspring will mainly determine its own characteristics, the offspring resulting from sexual reproduction may differ from one another. Small though such differences between individuals may be, they can determine different reactions to adverse conditions of the environment. Thus, a severe drought may kill off many plants of any one species, but enough will possess a sufficiently different inherited make-up that they are able to survive and perpetuate the race. In short, sexual reproduction plays an extremely important role in contributing to survival. However, some organisms do survive (and even *vary*) without sexual reproduction (and are said to reproduce asexually), but that is a long and complicated story not essential to our text.

Folklore, hearsay, and speculation – all have entered too often already into our consideration of tuckahoe. We should now, therefore, take time to examine in some detail the evidence upon which can be based our decision to accept or reject the statement that the hard stone which is the sclerotium of *Polyporus tuberaster* does indeed grow. Some publications on this subject are cited in the list of references given at the back of this book, and from these, one can easily check the details that have been reported about the further growth of the tuckahoe stone.

In the first place, one must possess a freshly dug tuckahoe or at least one which has not been allowed to dry completely. An old specimen preserved dry, let us say in a museum, cannot be expected to grow when moistened. Tuckahoes which have been reported to grow have been those that have been planted by the investigator soon after they were found or were received from the collector. The specimen has usually been planted about three inches deep in good soil in a pot which was then buried level with the ground in a garden or experimental plot, or in a woodland among poplars, and, of course, the planting must be kept moist. Specimens planted in May (in which month tuckahoes are often found) have been reported to produce the fruiting structure above ground in June or early July; more frequently it has been observed that fruit bodies are not produced until the following summer. Only one fruit body may be produced from one sclerotium, or sometimes several. Moreover, as we might expect, a storage organ as large as a tuckahoe may serve as a perennial structure; it has been reported that one tuckahoe planted in damp earth in a basement produced fruit bodies three years in succession.

The mushroom-like growths which develop from tuckahoes (Figure 19) vary from 2 to 6 inches in diameter and about the same in height. The cap of the fruiting structure is hemispherical at first but soon expands to a structure which is slightly curved on the

upper side and essentially flat on its lower side. This cap is attached to the sclerotium (from which the whole fruiting structure
grew) by a stout stem. The cap is light brown or buff in colour,
soft and cheese-like in texture, and is covered with minute buff-
coloured scales when mature. On the underside of the cap is a
layer of tubes, those near the stem end of the fungus being shorter
and shallower than elsewhere.

Because of the presence of the tube layer, a student of the fungi
is able to recognize that the mushroom-like body is a fruiting
structure representative of a group of fungi called the Polypora-
ceae, a group which includes many common and important tough
or woody shelf-like fungi commonly found growing on dead or
dying trees. Each pore of the pore layer is lined with basidia, very
similar to those of a mushroom, from which the reproductive
spores are shot off in tremendous numbers. The spores themselves
are microscopic and colourless but appear white in masses. Those
who know how to do it have been able to obtain cultures of tuckahoe from such spores, and it is reported that the mycelium which
grows from them is white at first but becomes pinkish buff and
then brown as it grows older.

Spores are discharged from the cap of the tuckahoe fruit, are
carried off by air currents, land on soil, and grow into mycelium
which in time will produce a new sclerotium or 'fungus store' in
the soil.

Published records also inform us that other kinds of similar
sclerotium-bearing fungi are to be found in other parts of the
world. It is now known, for example, that a fungus found in Europe, especially in Italy where it is called *'pietra fungaia,'* is the same
species as the Canadian Tuckahoe. The so-called 'blackfellow's
bread' of Australia, while similar in some respects, is, nevertheless, a species distinct from the Canadian one. One other sclerotium fungus of the same general sort (called the 'Southern
Tuckahoe') occurs in the southeastern United States; this species,

when it fruits, does not produce stalked mushroom-like bodies and is still another kind quite different from the Canadian Tucka-hoe.

The Ants'
Fungus Gardens
꩜

Although the welfare of mankind is involved with the lives of fungi in many ways and at many points in the life of each person, and often drastically complicated thereby, fungi do not constitute a large and important part of man's diet. It is true that many kinds among the larger fleshy fungi such as mushrooms, truffles, and morels are eaten, and it is also true that tons of mushrooms are grown annually (deliberately cultivated) for our tables; nevertheless, such fungi are in the nature of additions or luxuries, not staples of man's diet essential for survival. In fact, mankind would be hard put if there were nothing to eat except fungi. The more serious part of the food picture is of course that, in the past, fungi have sometimes destroyed crops utterly and caused widespread starvation; and they may do so even today.

But if man does not live by fungi alone, certain kinds of ants do. The attine ants which inhabit tropical and subtropical America depend for their welfare – for their very survival even – upon their ability to cultivate some kinds of fungi and upon their ingenuity and industry in achieving successful and continuous cultivation of their fungi.

The story of the truly astonishing relationship between the attine ants and their fungi seems more like science fiction than truth, even today when the results of many years of frequent painstaking observations and fully controlled laboratory experi-

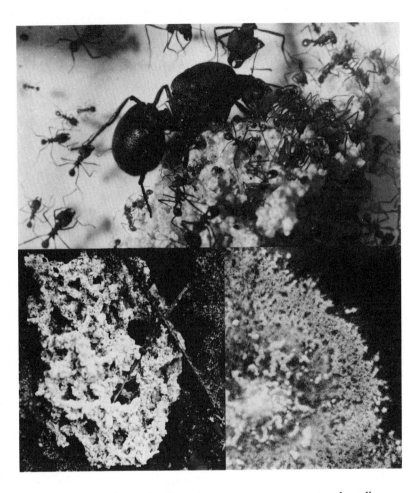

20 Fungus gardening by the attine ants: *upper,* queen ant and smaller
attendant workers, magnified about five times; *lower left,* part of an ant
garden (40 x 30 cm) growing in the soil of a moist ravine in Colombia;
lower right, the edge of a part of the ants' fungus as seen under the low
power of a microscope, at about 50 times natural size

The Ants' Fungus Gardens

ments are matters of printed record in words and corroborating photographs (Figure 20). Some early reports of the behaviour of the 'gardening ants' tended to be discredited on the grounds of lack of objectivity; but, for the most part, much if not all of what biologists exploring in the American tropics had reported as early as the middle of the nineteenth century has been confirmed by later investigators. A recent monograph on the topic lists about 300 articles and books dealing with it, and the bibliography is mostly readily available for any Thomas's who have the patience to search it out for reading.

The attine ants are also known as 'leaf cutters' because they forage, often far from their nests, in search of suitable green leaves which they cut into pieces and carry to the nest, holding large pieces over their heads and backs as might a man carrying a sheet of plywood. It has been established that many kinds of leaves are suitable for the ants' purpose, and it is reasonably certain that the attine ants are opportunists and help themselves most liberally to those kinds of leaves that are nearest at hand and most readily cut by their mouthparts. If easy-to-cut leaves happen to be plentiful and happen to be parts of such crop plants as cacao, coffee, citrus fruits, pomegranate, or other, so much the better for the queen of the colony of *Atta* ants and her enormous entourage of servants. Naturally, also, so much the worse for the unfortunate South American landowner who endeavours to earn a livelihood from those same crop plants. The destruction of tropical crops by the long soldierly lines of leaf-cutting ants forms the burden of many a heavy-hearted report sent home by the pioneers of the Spanish New World.

Biologists who first reported details about the ants' leaf-cutting activities assumed that the leaf pieces were used directly as food. By about 1874, however, it had become clear that the earlier assumption was incorrect and it is now known that leaf pieces are cut up and especially prepared in the nest to make a kind of compost upon which the ants are able to grow or cultivate certain minute fungi (Figure 20, both lower photographs). It is these fungi, or parts of them, that constitute the food for the ant colony.

One other aspect of the attine ants' harvesting operations puzzled observers. The actual labour involved is done mostly by fairly large 'worker' ants which, among the attines, are non-functional females (the queen ant being the only functional female who can lay fertile and viable eggs). Trotting along the forest path beside each worker there is usually a smaller worker who, it seemed to early observers, only added to the large worker's load because on the journey back to the nest the little fellows hitch a ride on top of the leaf piece. But it has turned out that the little ant hitchhikers provide compensation for *their* ride; they clear the leaf piece of unnecessary and undesirable bits of trash, they smear the leaf with digestive juice, and in other important respects they help to make the provender ready for the composting process which will take place in the nest.

Nests of attine ants (Figure 20) vary greatly in size and form. That part which can be seen above ground may be only a hole (often a crack in rotted wood) or it may be a low crater-like mound or even a kind of turret or chimney. A few kinds of ants make rather loose nests in rotted wood (Figure 20) well above ground, but more commonly most of the nest is below ground. If the nest is carefully excavated, it may be seen to be a veritable labyrinth of interconnected chambers, many of which are devoted to special purposes varying from middens or garbage dumps to brood chambers. Those who have studied the nests carefully have often remarked about the 'engineering skill' of the ants in ensuring that their home is provided with proper drainage, ventilation, and other desirable structural features.

As for the complex social life within the nest, much has been written in many books. Mostly this part of the story, fascinating though it is, is not essential for our consideration of the relationship between the ants and fungi. We may note briefly, however, that ants do not communicate with one another by means of any sounds transmitted through air. They produce vibration signals by means of their bodies (this is known as stridulation) and they are able to detect such vibrations and others transmitted through the substratum, that is, whatever solid object the ants are standing on at the time. Also, ants have very well developed compound eyes that serve for communication and in providing awareness of their environment. In addition, the tips of their antennae (or 'feelers') bear sense organs; the antennae are used much more than the mouthparts for detecting the suitability of the compost for their use and the degree of maturity of the fungal food.

If the above statements have aroused any interest then, at this point, there will be a torrent of questions, especially from the incredulous. Alas, in this story, the answers may strain credulity the more; for example, one of the first and most obvious questions will probably be: 'Where do the ants find the desirable fungi and how do they avoid fungi that are useless or perhaps even poisonous to them?'

In fact, the ants do not *find* the fungi which they use as food to sustain them, although at some time in the very remote past they must have found the fungi somehow. The fungus lifeline for the attine ants is their precious inheritance from their immediate ancestors; this is not an inheritance in some abstract form but in the form of an actual part or a subdivision of the living fungus which was cultivated by their progenitors and, farther back in time, by the progenitors of earlier attine ants.

The nucleus of the gardens of a newly founded colony is the 'dowry' of the queen ant. Before she goes for the nuptial flight, she places a quantity of the parental nest fungus in a special container or pouch called the infrabuccal pocket, which is a chamber located near the mouth opening. The mass of fungus and nutrient leaf bits *must* be carried with her and must be very carefully tended if a new colony is to be established successfully. After the mating flight, the queen ant drops to the ground and discards her wings as quickly as possible by frantically brushing them against dry grass or other brittle objects. The fungus 'culture' must not be allowed to dry to any great extent, otherwise it may not grow into a new garden. In dry weather and on dry soil, the queen ant immediately digs into the soil close to where she lost her wings. At one end of a short tunnel which she has made, she disgorges the precious pellet from her infrabuccal pocket and immediately lays several eggs upon and around it. Workers supply fresh leaf compost to establish a new fungus garden but if they do not act very quickly the queen herself may forage for the first leaf pieces required. Thus with purpose and great urgency (one is tempted to add 'with wisdom') is begun a new colony in which before long thousands of ants, their larvae, and adults of all sizes and classes will co-exist in a metropolis nourished (in a primary sense) by bits of green leaves on which will grow the fungus product of the queen's dowry (Figure 20).

Our fungal horticulturalists do not accomplish the building of a thriving colony, and so the perpetuation of their kind, without a truly Herculean effort. One can but try to imagine the 'hand labour' of excavation alone; millions of bits of humus and grains of sand must be carried out to create even a small part of the labryinth of chambers. Most kinds of ants do that kind of work, but it is the gardening labour of the attine ants that is so impressive. One almost feels exhausted when reading about it. Take first the matter of preparation of the compost, a matter so complex and demanding of devotion that one recent monographer (Weber, see bibliography) remarks: 'Even the compost beds created by [human] mushroom growers are simple by comparison'. When the fresh leaf pieces have been carried to the nest, other workers standing on or beside the garden cut the larger pieces up into much smaller fragments which are then chewed into a juicy mass

to which is added the ants' own feces. This latter item is of paramount importance, for the ants' feces contain certain substances
which encourage the growth of the fungus and, in addition, some
substances which break down or digest proteins and other food
substances present in the green leaves. Some kinds of ants also
add minute fragments of rotted wood to the compost and even
parts of the bodies of their deceased fellow ants.

Before a chamber is ready for a new garden, the entire floor of
that chamber is very thoroughly cleaned of every last bit of debris
of any sort. When new compost must be added to a garden, the
addition is made only to the uppermost part, or to the peripheral
parts. Although the ants weed their gardens in the sense of remov-
ing any extraneous material, it is believed that it is not the weed-
ing which enables the ants to maintain a pure growth of only one
kind of fungus; rather it is the addition of their own saliva and excrement which favours the growth supremacy of their particular
fungus over other moulds that might otherwise overgrow the ants'
special fungus.

One could easily indulge in loose thinking and characterize the
attine ants as 'gourmands' because they are so meticulous in the
details of ensuring the growth of an abundance of their fungus as
'a speciality of the house.' But there is ample evidence to establish
the statement that the ants' fungus is more than a speciality, it is
truly their staff of life; chemical analysis shows that it provides for
them a rich and complete diet.

It is now known that more than one kind of fungus can provide
food for the attine ants. When most of the fungi grow on the spe-
cially prepared compost, they take the form of a network of mycelial threads, essentially like the threads of many other kinds of
fungi. The fungi cultivated by the attine ants, however, produce
special structures known as staphylae. Each staphyla is a cluster of
hyphae (branches of the mycelial system) which are enormously
swollen at their ends. These globose or pear-shaped swellings con-

tain generous quantities of the food materials produced by the fungus from the leaf compost provided by the ants, and it is these staphylae which the ants prize and consume. With proper 'cropping' the ants maintain a constant supply of the swollen hyphal tips for their needs.

For many years, those who studied the ants' fungi were unable to classify them with certainty. In the first place, it was not a simple matter to induce the fungi to grow on the various laboratory decoctions (called media) which were devised in attempts to provide a substitute for the ants' compost. Further, some other kinds of fungi also produce swollen hyphae or mycelial tips. To identify the ants' fungi it was necessary to induce the fungi to grow in artificial culture and to induce them to develop their reproductive structures, for it is by study of the latter that fungi are most easily and certainly identified. Such difficulties were finally overcome, but not without many years of painstaking research. A few (not all) of the ants' fungi were obliging enough, after many 'diets' had been tried, to develop into various sorts of mushrooms and other closely related fungi that could then be identified. One kind turns out to be a species of *Lepiota* not far removed from certain mushrooms often found on lawns and in pastures; another (*Leucocoprinus*) is related to the Inky Cap Mushroom; still another is a kind of *Auricularia*, a genus that includes a jelly-like fungus that the Chinese people have long been able to grow for eating.

The secrets are out: we now know the names of the principal kinds of fungi that share their lives with the attine ants, but are we any the wiser for that? Not, certainly, about the intriguing problem of how such a complex and intimate association between the attine ants and their carefully tended fungus gardens evolved. From what is known of the mechanics of evolution and what is known about other kinds of ants' chance association with fungi, Dr Weber (see bibliography) made a few plausible suggestions but that is about where the whole matter stands.

To request any opinions from our blind men would only embarrass them, so remote would our story be from anything within their ken. Besides, they might feel inclined to philosophize about our evidence and begin talking about 'dedication to duty, the effect of labour unions in an ants' nest' and much more that is irrelevant. More likely one would say, very simply, '*only* a fungus in an ants' nest'; another would say, 'but surely an important lesson for mankind'.

Mushroom
Monstrosities

Malformed plants seem to have a strange fascination, owing partly perhaps simply to their deviation from the accustomed pattern but also owing partly to some additional special virtue which may more than compensate for their lack of conformity. A painter is often more inclined to select or emphasize an old weather-beaten and badly deformed pine tree or oak than to give prominence to an individual of regular or normal form. Possibly the artist subconsciously is paying homage to the strength and will to live that has enabled the pine to survive despite the adversity of winds, waves, lightning, and thirst. Such thoughts, however, are intangibles with which the biologist does not deal, not profitably anyway.

More tangible are considerations of what is known about the causes of malformation of plants. This is, by itself, a vast area of knowledge: though much is still to be learned, scores of books and articles have been written about it; at one time the subject became elevated to the status of an 'ology' under the dreadful sounding name teratology. Undaunted by that, it may nevertheless be of interest to give attention to a few simplified statements of what is involved.

One basic concept will assist us, and it is this. The 'normal' or usual form of a plant (or an animal) is fixed or determined within certain limits by the genes which it has inherited, most immediately from its parents, and cumulatively over a very long period from other individuals of its own kind or species which predated it. Of two plants which have almost identical form, one bears a rose which is white, the other red. Most of the genes possessed by each plant 'control' (or enable to be developed) those structural features by means of which we recognize each as a rose bush – such as the size and arrangement of the leaves, the saw-tooth edges of leaves, the surgically sharp thorns, the perfumed flowers. In one plant there is a gene (or possibly more than one gene) that enables red flower pigments to be produced, but no pigment is produced in the plant which lacks that gene. So profound may be the effect of a single gene that some plants of Indian Corn may be stalwart plants eight feet in height while others, which are otherwise identical in the matter of all their genes except one, are mere tufted bunches of leaves and no taller than two feet. First of all, then, marked deviation from the 'normal' or average form of any plant may be the result of the operation of a certain gene or genes which it possesses and which are not possessed by the average plant. We think of the extremely dwarf tufted corn plant as an abnormal form though it is really only a gene-induced variant of the usual type.

Genes, the determiners of inherited characteristics and the controllers of developmental patterns, not many years ago were little understood as to structure, chemical composition, and manner of operation. Today, genes are the focus of much exciting research which is probing into the innermost secrets of living material. Genes are now known to consist of long spirals of many complex chemical substances, the whole double spiral being rather like the two long pieces of a ladder with cross-struts; that is, roughly resembling a spirally twisted ladder. Moreover, it is now known that each gene serves as a highly specific template or model which has the property of ensuring that the particular proteins (such as enzymes) will have the proper chemical composition and conformation when these proteins are being put together from simpler pieces or units during growth of the plant. Geneticists now speak of the gene as possessing (or imparting) information. It is a word implying that the gene possesses the code necessary to ensure that all the complex processes both chemical and physical will take place, during growth and development, according to the pattern necessary if the cells of a plant are to develop as they should to become a rose or a cabbage.

Such an extreme simplification will not satisfy the curious, but it may be of some assistance towards realizing that the integrity of the genic, or heritable, material is of utmost importance to an organism. Damage of any kind or slight imperfection in duplication can result in an organism possessing genes that are no longer capable of ensuring that the required enzymes are produced – enzymes which are chemical agents which make possible a tremendous variety of the most basic chemical changes necessary for life. Thus damaged genes or 'incorrectly' duplicated genes are one cause of malformations in plants.

Another cause of abnormality or malformation is the presence of another organism, a parasite. A parasite may dwarf or stunt a plant simply by robbing it of its food. A parasite may also cause malformation of the plant on which it grows by producing chemical substances that affect basic aspects of growth. A virus, for example, may produce chemicals which either stimulate or retard the growth of the cells of the plant, especially in certain regions. One virus which affects the potato plant in such a manner causes the leaves to be extremely wrinkled instead of flat; another causes leaves to be highly mottled, the lighter areas intermingled with dark and containing less of the green leaf pigment, chlorophyll. Many common viruses cause the production of curiously flattened and often spirally twisted stems.

The object represented in Figure 21 is a malformed mushroom, the malformation being the result of parasitization not by a virus but by another fungus. Had the mushroom escaped being attacked by the parasite and had it developed normally, it would

113

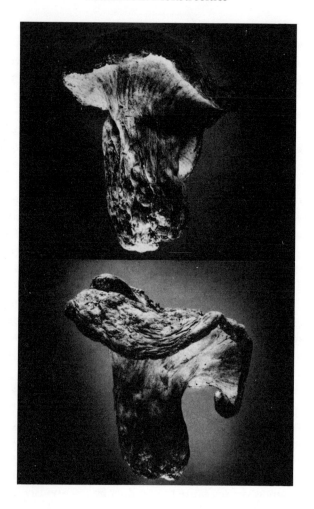

21 Malformation of a mushroom (*Lactarius*) is caused by a second fungus
(*Hypomyces*) growing parasitically on its outside. The orange-coloured
coating results in the irregular form and suppression of the spore-bearing
plates or lamellae. About four-fifths natural size

114

have appeared very much like the ones illustrated earlier in Figure 7 with a straight stem and a cap of regular form, from the underside of which thin spore-bearing plates would have hung down.

The diseased mushroom in Figure 21 is so malformed that the stem is no longer a stout regular cylinder, the cap bears little resemblance to the usual plano-convex form of most mushrooms, and the spore-bearing plates are almost non-existent or have been reduced to a few low and irregular ridges. This is only one example, but in fact mushrooms parasitized in this way assume a great variety of forms, from roughly spherical to highly lobed and contorted.

Like many malformations, however, diseased mushrooms possess a compensating characteristic which often results in their attracting much attention. Not only is their form so bizarre that one wonders what manner of 'growth' they can be, but also their colour is so vivid that they immediately catch the eye. Some are bright yellow-orange, or orange-red, some are scarlet and some purple-red; an occasional specimen may even exhibit all those colours on different parts. If the mushroom which has been parasitized happens to be, in its normal or unparasitized form, a rather brightly coloured one, some of the show of colour in the malformation may be the residue of the original colour. But frequently the victim is a rather drab or weakly coloured species, the flaming pigment in such cases being attributable to the parasite.

The almost complete obliteration of spore-bearing plates or lamellae as a result of malformation causes the diseased mushroom to be sterile because no basidia or spores are formed. Some individual mushrooms (of a kind that harbours the parasite) always escape infection and it is such healthy specimens which do reproduce. In this way there is always a supply of the host mushroom somewhere for the parasite to live upon. Mushrooms most commonly attacked are certain species of the genus *Lactarius*, a group characterized by the production of a sticky milky juice which may be seen to ooze out from their tissues when such mushrooms are broken.

The parasitized mushrooms have often been reported to be edible. To use them as food, however, would be extremely dangerous; because of the extensive malformation and lack of spores it is usually impossible to identify the kind or species of mushroom attacked. If it was a poisonous species before it was parasitized, it remains poisonous and thus the malformed parasitized mushrooms should never be recommended as food.

And where is this parasite that brings about malformation and what is it 'like,' as the blind men would say? If we can escape for a moment from the distraction of the sheer pleasure evoked by the lovely orange colour and begin examining the mushroom closely, even with the unaided eye, we perceive that the surface is bumpy or pimply, especially on the lower side of the cap where normally the spore-plates would be present. Because we have not previously seen mushrooms with such a conspicuously lumpy surface, we conclude that the lumps must be the parasite, or part of it. When we examine this lumpy surface using a magnifying glass so that we are seeing it ten times enlarged, the image might suggest fine sand in which there are embedded, at regular intervals, a number of pears which protrude from the sand by only a third or less of their length. Slightly higher magnification, as would be provided by a stereoscopic microscope, would reveal something else to whet curiosity: the necks of the pear-shaped bodies are open – anyway, some of them are – as though the stem pieces had been pulled out of the necks of the microscopic pears.

To learn more about the fungus parasite it is now necessary to cut very thin slices vertically through the parasite and the mushroom tissue in which the parasite is embedded. When such slices are then examined by using a microscope, it becomes clear that there are really two regions of fungus tissue: one which is the mushroom underneath, and another which covers or seems to be plastered all over the outside of the mushroom and made up of filaments which are more densely compacted than those making up the mushroom. This dense plastered-on layer is the stroma or basal ground work part of the parasite. Furthermore, most if not all the flaming colour is present in the stroma; the diseased mushroom has taken on a rather thick orange mantle which covers it almost wholly.

What appeared to be lumps or pear-shaped bodies can now be recognized as fungus fruiting structures called *perithecia*. Each is embedded partly in the stroma except for the neck which projects beyond it, and each neck is open to the outside because of a channel through it. Arising as a cluster from the inside at the base of each 'pear-ish' perithecium is a cluster of long cylindrical tubes and, looking carefully we realize that each tube bears within it a number of long thread-like slightly curved spores, the walls of which are delicately rough. The long tubes are asci which contain ascospores; and the production of spores (usually eight for each ascus) in a sort of pea-shooter tube is the basic feature of a very large group of fungi named the Ascomycetes. A bit of hunting through the appropriate manuals indicates that our parasite is none other than *Hypomyces lactifluorum*. A mushroom which produces its spores in groups of four, each group surmounting, on the outside, a club-shaped basidium, is classified as one of the Basidiomycetes. The orange overlayer is a fungus which produces eight spores inside a long tube or ascus and is classified as one of the Ascomycetes. The two classes of fungi are very far apart in the classification scheme and there are large numbers of basic differences between the two groups. But there are also examples of one kind of mushroom existing parasitically upon another kind of mushroom in much the same kind of relationship as that of our present examples – a fungus that grows upon a fungus. Does it remind us slightly of the 'greater fleas and lesser fleas'? And what of beauty? That there is surely in the vivid orange colour; beauty of form there may be also, depending on how we see our fungus. Notice how different the one malformed mushroom specimen appears in our two photographs (Figure 21), where only the position of the camera represents a different viewpoint.

Is it ridiculous to speculate as to what the one blind man would have given as his opinion about the elephant's leg if it had been covered on the outside by soft plaster? I believe he might have done exactly what we have been doing all along: he just would have dug a little deeper.

Postscript

In these pages I have striven to record only a very small part of what is known about that great and numerous tribe of man's fellow creatures, the fungi. I have sought to recognize truthfully their 'greatness' as seen from man's viewpoint, whether measured by destruction of crops by Rust Fungi or measured by saving of lives through the virtue of vital chemical processes of the mould fungus *Penicillium*; but I have sought, also, to emphasize the greatness (and that in their own right) that is inherent in the fascination of the almost limitless forms and colours of fungi, from which they remain for man ever a source of the 'delight of his curiosity,' as Linnaeus wrote in 1752.

Such claims I have made for the adjective 'great.' And 'numerous' applied to the fungi is unquestionably a conservative adjective. Biologists have given names to what are supposedly recognizable separate entities to the extent that about 100,000 kinds or species are known. Kinds or species, that is; but when one deals with the numbers of individuals of any one species that may inhabit the minutest corner of the earth, one needs to invoke astronomical numbers or else employ the mathematicians' concept – namely infinity. How many individuals of the microscopic bread yeast fungus, for example, are present in a cup of active fermenting yeast?

In recording a few facts pertaining to a few chosen fungi less familiar than mushrooms, puff-balls, and the like, I have been aware of having allowed myself only one major latitude, though I may have been guilty of several or even of laxity. That licence, if such it is, has been dictated by conviction that the beauty of natural objects should be constantly drawn to our attention and emphasized whenever possible. But other than mutely to point and to hope that others can share, what else remains when one is confronted with the beauty of a living Coral Fungus?

There are things one can do in order to communicate with others about a Coral Fungus; of these, what needs to be written or said depends upon the viewpoint or the purpose underlying the desire to communicate. Thus, one could begin with what might appear to be the simplest sort of communication, a description; the form and wording even of this, however, will depend upon who writes it and to whom it is directed. A biologist writing a description of a species which has never been found or named before is obliged to use terms which are as precise as possible in the sense of being comprehensible to the reader, yet not subject to variation in the matter of interpretation by other biologists who will read the description. For example, if the colour of the fungus is 'fawn,' that adjective would be perfectly satisfactory for many purposes, such as in a literary context. But because it is subject to a broad interpretation, the biologist would feel obliged to search for a more precise term such as one which could be found in one of several manuals in which a great variety of shades of all basic colours are denoted by standard names, or even represented by numbers.

Perhaps the example is sufficient to support the concept that precise unequivocal writing (especially in science, where observations and facts must be stated with as high a degree of objectivity as possible) tends to result in pages that, often, do not make the most attractive reading. The scientist has been schooled in the suppression of words that reflect emotion or that in any other way might cast doubt upon his ability to record factual information accurately or upon his ability to draw logical conclusions from those facts. In other words, the simple unadorned statement of facts as they are set down in the lengthy and extensive records of biological science frequently does not provide a good medium through which a biologist is able to share with others much of the excitement or delight which he may feel personally.

A fact is something known by observation or experience to be true or real; however, something asserted to be true or to have happened is also frequently accepted as a fact. Therefore, because even our generally used concept of the word 'fact' is seldom defined unequivocally, facts are too often quoted as the ultimate or determining reference or bench-mark or else are regarded as something to be eschewed or ignored. In science, facts are accepted as the most reliable medium for communicating exactly the concepts of one mind to another mind – no more.

Though facts may not appeal to the aesthetic sense or to literary taste, some degree of awareness of them can, nevertheless, add to our appreciation of natural objects with which we may have been unfamiliar. The unusual and elegant form of the microscopic mould *Pilobolus* is so attractive in itself that a news reporter (who may have been one of the blind men) felt compelled to write: 'a beautiful photograph of some curious objects made of blown glass.' Beauty there is indeed to charm the beholder, but how much more intriguing *Pilobolus* becomes for us if we can also appreciate the beauty of the astonishing manner in which the fungus carries out the essential function of spore dispersal for survival! Aware of even a few of the hard facts about *Pilobolus*, we find ourselves more deeply interested in it than before and realize the significance of even so lowly an organism in relation to other entities of the living world.

Although I have tried to respect your feelings in the matter, I have here and there uttered a plaint about the need for restraint in the expression of questions involving the 'good' or the 'use' of fungi. Among the examples recorded in these pages, the use of *Penicillium notatum* is unlikely to be contested because its precious drug, penicillin, can be used to save human life. But the use of the sphere thrower *Sphaerobolus* is another matter. As far as is known at present, that wonderful little fungus mortar lives its life unheralded merely because it happens not to have any direct connection with or effect upon human welfare: for me, it nevertheless remains an object of beauty and a source of satisfying mental stimulation; for me, the good or use of any living thing does not depend solely upon whether or not it happens to fit into man's limited viewpoint. As far as is known at present, the fungus that provides livelihood for the ants that grow it as a crop is of no use to mankind, but it is entirely possible that some day the ants' fungus could serve mankind. Meanwhile we can speculate on what we might learn from the ants' viewpoint if they could explain it to us, which, regretfully, they cannot do.

In retrospect, I would conclude that the three blind men do not know very much about fungi after all – about the kinds we have inspected, anyway; a fungus is a fungus, and many fungi seem not to be much 'like' anything else. Yet, even when dealing with fungal oddities, the message of the blind men is clear: our evaluation of the fungi as repugnant or as beautiful, fascinating or uninteresting, useful or useless, is in large measure based upon impressibility and viewpoint. The influence of a natural object such as a fungus depends upon the individual who sees the fungus!

Glossary

algae Flowerless chlorophyll-containing plants of simple structure; mostly aquatic; including seaweeds and pond scums.

alkaloid An organic substance having alkaline properties and occurring naturally in plants and animals; alkaloids are composed of nitrogen, carbon, and hydrogen (a few have oxygen as well); some are important drugs.

ascospore A fungus spore produced following a sexual process and developed within a microscopic sac-like structure which is often tubular.

ascus A microscopic sac-like structure produced by certain fungi (called Ascomycetes), enclosing several spores (usually eight) and developing as the result of a sexual process.

attine (ants) Tropical and subtropical ants which cultivate fungi as food. Named for one of the principal types, the genus *Atta*.

bacterium A microscopic unicellular organism which has its genetic material dispersed in clumps rather than located in an organized nucleus.

basidium A microscopic club-shaped structure on the outside of which are attached several spores (usually four) and developing as the result of a sexual process (see *ascus*).

basidiospore A fungus spore produced following a sexual process and developed on the outside of a basidium (q.v.).

carbohydrate An organic compound consisting of a chain of atoms of carbon to which atoms of oxygen and hydrogen are attached; examples are sugar and starch.

cellulose A complex carbohydrate (q.v.) which constitutes much of the hard structural material of higher plants.

colloid Any glue-like substance (gelatin, starch, raw egg-white, etc.) that does not diffuse through plant and animal membranes or diffuses only very slowly.

compost A mixture of decomposed vegetable matter used to promote growth of garden plants.

culture The growing of living organisms on an artificial 'medium,' for example, on a jelly containing nutritional chemicals.

cytoplasm The living matter of a plant or animal cell exclusive of the cell wall or cell membrane and the nucleus; mostly a complex mass of proteins.

eel-worms Minute worm-shaped animals (*Nematoda*) present in soil, some of which infect roots.

enzyme A protein (q.v.) capable of activating a particular chemical reaction.

ergot The sclerotium or resting stage of a fungus which parasitizes rye and other grasses.

fungus A plant which lacks the green pigment chlorophyll of flowering plants and whose body form is a thallus (q.v.) not bearing recognizable leaves, stems, and roots but which most commonly has the form of branched microscopic filaments. Fungi usually reproduce by means of sexually or asexually produced spores.

gamete A specialized sex cell (or a sex nucleus) whose function is to unite with another in sexual reproduction.

gene A repository of a unit of genetic information necessary for the particular organism.

genetics The science which deals with heredity.

gel The partly solid state of a colloidal system.

glycogen A complex carbohydrate consisting of a large number of molecules of sugar (glucose) chemically united.

hormone A chemical substance produced in one part of an organism and affecting another part of that organism.

humus The organic part of soil (e.g., decomposed plant material) as opposed to the inorganic part (e.g., sand).

hypha A tubular filament; a unit of the mycelium (q.v.) of fungi.

lamella A thin plate-like structure on which some fungi produce spore-bearing basidia (q.v.).

lignin A complex chemical which makes up a large proportion of the substance of wood.

Glossary

liverwort A prostrate moss-like flowerless plant whose body often has the form of ribbon-like branches.

mitochondria Grains or particles of the cytoplasm (q.v.) which serve as the site of the chemical energy-producing process known as respiration.

mycelium A system of branching microscopic tubules constituting the body of most fungi.

mycology The science concerning fungi.

nematode A class of small roundworms.

nucleus The central and 'governing' body of all living cells with the exception of certain very primitive kinds (*Monera*).

osmosis The process in which water migrates through a semi-permeable membrane.

parasite An organism living upon or within another organism (host) and to some degree detrimental to the host.

perithecium A minute spherical or pear-shaped fruiting structure of certain fungi, often opening by an apical pore.

protein A complex organic compound formed of smaller units (amino acids) and constituting a large proportion of living matter, protoplasm (q.v.).

protoplasm Living matter or substance.

protozoa Microscopic single-celled organisms, largely aquatic.

ribosome A cytoplasmic granule which is the site of protein synthesis in living cells.

sclerotium A hard mass of fungal tissue containing a reserve of food material.

sol The liquid state of a colloidal system.

sporangium A microscopic asexual spore-producing and spore-containing structure.

spore A microscopic cell capable of developing into an adult organism.

stipe A stalk-like support or stem.

substrate An underlying layer.

thallophyte An older term applied to flowerless plants not differentiated into root, stem, and leaf.

thallus A plant body not differentiated into root, stem, and leaf.

trichogyne An elongated projection from the female structure which is receptive to the male gamete.

vacuole A small space within the cytoplasm (q.v.) and containing non-living material.

Selected References

Alexopoulos, C.J. *Introductory Mycology*, 2nd ed. (New York: John Wiley and Sons 1962) 613 pp

Barron, G.L. *The Nematode Destroying Fungi* (Guelph: Canadian Biological Publishers Ltd. 1977) 140 pp

Brodie, Harold J. *The Bird's Nest Fungi* (Toronto: University of Toronto Press 1975) 214 pp

Christensen, Clyde C. *The Molds and Man* (Minneapolis: University of Minnesota Press, Minneapolis 1972) 284 pp

Crowder, W. 'Marvels of Mycetozoa,' *National Geographic Magazine, 49* (1926): 421-3

Duddington, C.L. *The Friendly Fungi* (London: Faber & Faber 1957) 188 pp

Gray, W.D. *The Relation of Fungi to Human Affairs* (New York: Henry Holt & Co. 1959) 510 pp

Groves, J.W. *Edible and Poisonous Mushrooms of Canada* (Ottawa: Canadian Department of Agriculture 1962) 298 pp

Hard, M.E. *Mushrooms, Edible and Otherwise* (New York: Hafner 1961) 609 pp

Ingold, C.T. *Dispersal in Fungi* (Oxford: Clarendon Press 1953) 197 pp

Large, E.C. *The Advance of the Fungi* (New York: Henry Holt & Co. 1940) 488 pp

Miller, O. *Mushrooms of North America* (New York: E.P. Dutton & Co. 1972) 360 pp

Ramsbottom, J. *Mushrooms and Toadstools* (London: Collins 1953) 306 pp

Rolfe, R.T., and R.W. Rolfe. *The Romance of the Fungus World* (Philadelphia: J.B. Lippincott Co. 1928) 309 pp

Selected References

Shanor, L. 'Some Observations and Comments on the Laboulbeniales,' *Mycologia*, 47 (1955): 1-12

Tosco, U. *The World of Mushrooms* (London: Orbis Publishing Co. 1973) 128 pp

Vanterpool, T.C., and Ruth Macrae. 'Notes on the Canadian Tuckahoe,' *Canadian Journal of Botany, 29* (1951): 147-57

Weber, Neal A. *Gardening Ants: The Attines* (Philadelphia: American Philosophical Society, Memoirs, Vol. 92, 1972) 146 pp

Index

Index

Index